Inclusion by Design

This book introduces a new speculative design process for inclusive new product development (NPD). The authors offer Vision Enabled Design Thinking (VEDT), a human-centered technological design framework incorporating the use of Design Lens and Vision Concepting, as a way for the designer to ideate and reflect on product development concepts within a deeper sociocultural context. The authors incorporate project management concepts into the overall design process through the development of a new design process, 4-D Algorithm for New Product Development.

Inclusion by Design: Future Thinking Approaches to New Product Development formalizes the use of speculative design as a means for more inclusive NPD and promotes management of the design process as a needed skill for future engineers and designers. It provides a novel design methodology of VEDT for engaging vision concepting, through the use of design lenses in engaging speculative design practices and offers an implementation framework to support the sustainable adoption and use of future design methods. The 4-D Algorithm for New Product Development promotes inclusivity in design while addressing practical aspects of managing the design process in today's corporate business environment.

Those involved with interactive product and technology design, new product development, design researchers and managers, engineers, as well as professionals and graduate students will find this book useful.

T0351146

Inclusion by Design

Future Thinking Approaches to
New Product Development

Frances Alston and Emily Millikin DeKerchove

CRC Press
Taylor & Francis Group
Boca Raton   London

CRC Press is an imprint of the
Taylor & Francis Group, an **informa** business

Designed cover image: Shutterstock

First edition published 2024
by CRC Press
6000 Broken Sound Parkway NW, Suite 300, Boca Raton, FL 33487–2742

and by CRC Press
4 Park Square, Milton Park, Abingdon, Oxon, OX14 4RN

CRC Press is an imprint of Taylor & Francis Group, LLC

© 2024 Frances Alston and Emily Millikin DeKerchove

Library of Congress Cataloging-in-Publication Data
Names: Alston, Frances (Industrial engineer), author. | DeKerchove, Emily
    Millikin, author.
Title: Inclusion by design : future thinking approaches to new product
    development / Frances Alston, Emily Millikin DeKerchove.
Description: First edition. | Boca Raton : CRC Press, 2024. |
    Includes bibliographical references and index.
Identifiers: LCCN 2023003362 (print) | LCCN 2023003363 (ebook) |
    ISBN 9780367416874 (hbk) | ISBN 9781032537597 (pbk) |
    ISBN 9780367854720 (ebk)
Subjects: LCSH: Product design. | New products.
Classification: LCC TS171 .A65 2024 (print) | LCC TS171 (ebook) |
    DDC 744.7—dc23/eng/20230227
LC record available at https://lccn.loc.gov/2023003362
LC ebook record available at https://lccn.loc.gov/2023003363

ISBN: 978-0-367-41687-4 (hbk)
ISBN: 978-1-032-53759-7 (pbk)
ISBN: 978-0-367-85472-0 (ebk)

DOI: 10.1201/9780367854720

Typeset in Times
by Apex CoVantage, LLC

Contents

PART 2   How Does Organizational and Individual Readiness Support Inclusivity into the Design Process?

PART 3   Case Study: Grand XI Personal Fitness and Wellness Coach

Preface

In meeting the new product development (NPD) challenges of the 21st century, the engineer as a technologist and product designer must not only grapple with technological considerations, but also the sociocultural implications of their designs. Unfortunately, often as a function of the depoliticization of engineering education, the engineer is often methodologically ill-equipped to appropriately understand and address these non-technical concerns. This narrowed perspective can cultivate exclusionary engineering design and analogous decision-making that can lead to biased technological systems that are devoid of the needs and considerations of sociocultural diverse user groups.

To expand the bounds of the designer's questioning, thus, affording greater consequential and inclusive design thinking, "design thinking must be futures empowered" asserts Frank Spencer, Founding Principal and Creative Director of the Futures School. Futures thinking helps to **"empathize, define, ideate, prototype and test"** in a much more holistic, emergent and transformational way than is possible without it." As there remains limited available academic research in exploring futures thinking in design, "how" remains the question. This book, in response, proposes an engineering management pathway forward for academic classroom considerations.

This book introduces and describes a newly developed design model (4-D Algorithm for New Product Design) focused on human-centered technological design framework that incorporates the use of design lens and vision concepting, a speculative design technique, as a means for the product and technology designer to both ideate and reflect on technology concepts within a deeper and richer sociocultural context. This human-centered process is expected to afford a critical reflection that will broaden and deepen the design engagement, fostering, ultimately, more inclusive and innovative technological solutions. In advancing the practice and discipline of engineering management, this book also discusses the organizational design and deployment considerations in enabling inclusive approaches to NPD, such as Vision Concepting. As recognized, the engineer or designer is often responsible for not only the design of a new or updated product, but also overall management of a team or organization. A specific emphasis is made in identifying tools and practices that support individual, team, and organizational readiness and design within the technology enterprise.

This book has been organized into three parts: Part 1 discusses different approaches to design thinking, such as humanistic, futuristic, and speculative design, and defines where inclusivity factors (Design Lens) are identified and incorporated into a new design process (Vision Concepting and 4-D Algorithm for New Product Design). Part 2 discusses leadership, organizational, and individual readiness required to support effective implementation of the design process and presents a framework for individual thinking and for educating and training new design engineers to facilitate inclusive designs. Part 3 presents a case study which incorporates inclusivity and organizational and individual readiness through the application of Design Lens, Vision Concepting, and the 4-D Algorithm for New Product Design.

Through application of the 4-D Algorithm for New Produce Design, in concert with Design Lens and Vision Concepting, the engineer or designer will not only instill inclusivity into the overall design process but will also increase the number of customers that could use a product which leads to increased corporate profitability.

BIBLIOGRAPHY

Cech, Erin A., and Heidi M. Sherick. "Depoliticization and the structure of engineering education." International perspectives on engineering education. Springer, Cham, 2015. 203–216.

Evans, Martyn. *Design Futures: An Investigation into the Role of Futures Thinking in Design.* Lancaster University (United Kingdom), 2010.

Polakova, S., Cruchade, A., "Can Speculative Design Make UX Better?" (design trend 4/4), October 24, 2018 https://uxplanet.org/can-speculative-design-make-ux-better-design-trend-4-4-ce8d13148e5d

Author Biographies

Emily Millikin DeKerchove

Emily Millikin DeKerchove has over 38 years of leadership experience in regulatory, environmental, radiation protection, and safety and health at Department of Energy (DOE) and Department of Defense (DOD) chemical and radiological operations and remediation. She has held various executive leadership positions in different operational environments and is a proven leader in achieving excellence in both program and field execution of safety and health, radiation protection, quality assurance, environmental, industrial hygiene, safety culture, and voluntary protection programs. Ms. Millikin DeKerchove earned a B.S. in Environmental Health with majors in both Industrial Hygiene and Health Physics from Purdue University. She is a certified industrial hygienist (CIH) and certified safety professional (CSP) and serves as an advisory board member to the DOE National Supplementary Screening Program.

Dr. Frances Alston

Dr. Frances Alston is an instructor at a major university and the past president (2018) of the American Society for Engineering Management. She is also a fellow of the American Society for Engineering Management (ASEM) and has built a solid career, leading the development and management of environment, safety, health, and quality (ESH&Q) programs in diverse cultural environments. She has a Ph.D. in Industrial and System Engineering and a MSE degree in Engineering Management, both from the University of Alabama. She earned a master's degree in Hazardous and Waste Materials Management/Environmental Engineering from Southern Methodist University and a Bachelor of Science Degree in Industrial Hygiene and Safety/Chemistry from Saint Augustine's University. She holds certifications as a certified hazardous materials manager (CHMM) and a certified professional engineering manager (CPEM).

Part 1

How Does a Product Designer Incorporate Inclusivity Into Their Design Process?

1 Inclusion by Design

1.1 DESIGN THINKING

Design thinking is a methodology used to develop design concepts and includes application of the concepts to a product or service. Design thinking primarily originated in the 20th century and has been applied in many ways to various entities, including business management and new product development. The initiation of design thinking came into existence as a method to improve the design process by considering various factors or variables which can influence the acceptance of a design. One of the early recognized experts of design thinking, Peter Rowe, author of, *Design Thinking,* applied the concept of design thinking into design for architecture using conceptual approaches to design in terms of both process and form. Over time design thinking has evolved to include various design processes developed to improve and enhance new or existing designs. Design thinking can also be thought of as being centered on human beings, similar to human factors, by understanding the customers' needs at the beginning and then including those needs throughout the development, testing, and deployment process. Design thinking techniques can also be applied to business problems and assist leaders across industries rethink their products, expand their market base, offer greater value to customers, and innovate to stay relevant.

The application of design thinking has been recognized as an innovative approach to business management by Tim Brown, who used integrative thinking and the ability to balance human feedback with design of business organizations and teams. Design thinking generates a creative resolution, and the model contains elements of individual processes to create the best one. Through the application of design thinking, Tim Brown has been successful in applying the methodology to successful businesses who are wanting to strengthen and improve on innovation of their brands. One of the more common design thinking models is the Double Diamond Design Process, as developed by the United Kingdom (UK) Design Council.

1.2 INCLUSIVE ENGINEERING DESIGN

The products, services, and technologies that are helpful in enhancing culturally and socially the quality of life have a distinct role in shaping the overall culture present and into the future as incorporated into design thinking.

"As technology becomes more ubiquitous, it is essential we consider the impacts on people, whether unintended consequences or designs that exclude certain groups or disadvantage them in some way" Mark Searle, executive vice president of Arizona State in the New York Times article, "Top Universities Join to Push 'Public Interest Technology.'"

Demonstrating an understanding of consumers' needs increases the likelihood that the final designed product will meet the needs of people. Learning to understand

DOI: 10.1201/9780367854720-2

users and their experiences requires research and deliberate decision-making. Incorporating inclusivity into design requires designers and engineers to understand the users' behaviors, attitudes, and characteristics of the tasks that will be performed while using the product. There have been significant advances in facilitating the design of products and services that satisfy the needs of different users and are easy to use; however, some design organizations still fail to consider and acknowledge users' needs early on in the design process. Inclusive design, a user-centered design approach that can be applied when design thinking, can assist designers expand the boundaries of product usage for as many people as possible by repeatedly adjusting product design to the needs of myriad of users from the start of the design process.

Support for inclusion in design continues to grow as various leaders and practitioners continue to voice their support. As Salima Bhimani, Chief Equity and Inclusion Strategist at Other Bets at Alphabet Inc., in a Google Blog piece entitled Product Inclusion Leadership, supports the case for inclusion in design. She states:

> we are at a pivotal moment in understanding that when we build technologies and companies that reflect the knowledge, skills and needs of those who are often outside the center of design, power and impact—such as women, people of color, people with disabilities and others—we see exponential benefits. Benefits to what our businesses can achieve, innovatively, and benefits in shaping generational experiences and realities—and great monetary benefits.

Inclusive design is a strategic approach to designing products, services, and environments and takes into consideration the needs of the largest number of people as possible. It is most frequently used to understand marginalized, overlooked, or vulnerable populations. Accessibility and equal opportunities for all in the information age have become increasingly important over the past decade. As such, design for all, universal access, and inclusive design are some of the different terms for approaches that are focusing primarily on increasing the accessibility of systems and processes for the greatest range of use. Some inclusive design approaches are detailed in Table 1.1.

The various approaches to designing for accessibility closely aligned with inclusivity as part of design thinking. Upon review and analysis of the various design methods, definitions, and method purpose, several themes emerge that underpin the concept of inclusive design as part of design thinking. These themes include

1. Inclusive design is the result of intentionality in action; a methodology that is engaged throughout design and development.
2. Inclusive design draws on the diversity of humanity in informing design decisions.
3. Inclusive design is designing for a diversity of ways and people so that everyone has a sense of belonging.
4. Inclusive design is an extension of excellence in design.

The four themes listed above were further broken down into three tenets that are listed in Table 1.2 and are discussed in further detail in Chapter 10.

TABLE 1.1

Design Thinking Evolution for Inclusivity

Design Approach	Definition
Design for All	Design for human diversity, social inclusion, and equality.
Universal Design	The design of products and environments to be used by all people to the greatest extent possible, without the need for adaptation or specialized design.
Inclusive Design	The design of mainstream products and/or services that are accessible to, and usable by, as many people as reasonably possible on a global basis, in a wide variety of situations and to the greatest extent possible without the need for special adaptation or specialized design.
Accessible Design	No individual shall be discriminated against on the basis of disability in the full and equal enjoyment of the goods, services, facilities, privileges, advantages, or accommodations of any place of public accommodation by any private entity who owns, leases (or leases to), or operates a place of public accommodation.
Cooperative Design/ Participatory Design	A design process involving full cooperation and participation between the users and the development team, who share their knowledge and experiences while designing together.

TABLE 1.2

Tenets of Inclusive Design

Tenet	Contribution
Consequential	A design methodology that is engaged throughout the design and development and results from intentionality.
Culturally Calibrated	Designs are focused on unique characteristics of diversity, so everyone has a sense of belonging and inclusion.
Equitable	Design draws on the diversity of humanity in informing design, in particular, those typically underrepresented and marginalized (those often excluded).

These tenets are key for developing and launching new products and technologies that can be embraced by a diverse consumer base. The case for inclusive design has been made and continues to be made by practitioners.; it is only a question of how that remains to be determined.

1.3 INCLUSION BY DESIGN MAKES GOOD BUSINESS SENSE

Companies are always looking for ways to expand their customer base. The primary purpose of their survival is to grow profits while providing a quality product; such that companies are continuously looking for ways to reach more customers across the world and to become the dominant supplier. One need to only look at McDonald's, Starbucks, or Apple who have stores all over the world. Accessibility to a sustainable Wi-Fi signal and use of the internet to purchase products have exploded the ability of companies to sell their products in many countries not previously accessible.

The ability to understand how a customer thinks and what products they consider to be important in their life, can have a significant impact on how well the product is received by the consumer. Applying design thinking and inclusive design, into development or redesign of a product allows the designer to strategically incorporate inclusivity into the design process. As recognized by Bruno Perez, there are four key methods one can use inclusive design for business advantage:

1. Customer engagement and contribution. "Engagement with a product increase when it is easy to use, and the experience is inclusive for most people. The key to this business justification is to demonstrate exactly how mismatched designs are affecting our customers."
2. Growing a larger customer base. "The strength of this approach is that it outlines very clear constraints, helping the team build a deep understanding of the best way to connect with a wider target audience, based on their context instead of their abilities."
3. Innovation and differentiation. "Inclusive solutions, in particular, have a history of seeding innovation that goes on to benefit a wider audience—new ingredients generate new products. A shift in perspective and context can also lead to new usage patterns and purpose to a solution."
4. Avoid retrofitting. "Many teams and companies treat inclusion as an add-on, something to consider only in the final stages of completing a product. When a solution is treated as 'for disabled' or 'accessible' there is often little, or no attention paid to the design as we keep considering it an add-on."

There are additional benefits to applying design thinking, and inclusive design, into the design process. A great example is name recognition. As a company's customer base expands, so too does the name recognition of the product. A great example of this is Apple. As the internet grew, so too did the company and name recognition of Apple and their products. As recognized by the BBC, there are five things that make Apple a recognized corporate leader:

- Steve Jobs was a brand unto himself.
- The iPhone was a revolution.
- Apple services and brand loyalty
- China and the business' growth—which represents a quarter of Apple's profits
- The Apple brand itself

The name recognition of the product itself, which can develop into brand loyalty, is as successful as sales of their products, and as their business continues to grow, the application of inclusive design will continue to be factored into their products.

1.4 MEETING THE SOCIOCULTURAL CHALLENGES OF PRODUCT DEVELOPMENT

In meeting the new technology-based product development challenges of the 21st century (e.g., artificial intelligence (AI) products and technologies, autonomous vehicles) the designer and engineer must not only include and manage technical considerations into the design process, but also the sociocultural consequences of their decisions that lead their design practices and design outcomes. The sociocultural challenges of current day present real issues for designers and engineers in developing products and technologies because of the wide array of factors that must be considered for inclusivity and usage. Sociocultural factors of considerations include all aspects of the way people live, their values, and their customs. These factors can include but are not limited to culture, language, customer preference, education, and religion.

Sociocultural theory is known as a theory of the mind that focuses on the relationship between the psychological aspects and the artifacts associated with transforming and guiding cognitive and mental functions. The products, services, and technologies that are prominent in enhancing the quality of life culturally and socially can lead to important cultural artifacts that have a role in shaping the overall culture present and into the future. Because of the relationship between artifacts and their ability to transform or guide cognitive and mental functions, it is incumbent upon designers to understand the impact of their design on humans.

When framing thoughts on intentionality in design decision-making, there is a call to ensure with reasonableness that designs are approached with inclusivity, taking into considerations the sociocultural aspects. This begins with designers that are focused on assembling a diverse team in an effort to achieve differences of thought and experiences. In some workplaces it is not easy or even achievable to charter a diverse team due to the makeup of the organization. When a diverse team is formed, active participation of each member with different experiences, thoughts, and ideals is preferred over passive participation to yield the ideal design and subsequently products. Active participation increases the opportunity for inclusivity of team participants and development of products that can be used safely that will perform at its maximum potential for a diverse cadre of users. There have been occasions reported where newly developed products and technologies could not be effectively used by a diverse consumer base. The flaws in product usage may have resulted from overlooking some sociocultural factor that have limited the effective usage of a product by some.

In the span of two days in November 2019—November 19, 2019, to November 20, 2019—the *New York Times* published two informative articles: "Dealing with Bias in Artificial Intelligence" and "The Big Business of Unconscious Bias" which was related to addressing bias in technological design. Seemingly these articles may be coming from differing viewpoints on bias; however, when considered in combination, the articles, together, elucidate the challenge of inclusive technological design, demonstrating a need for greater intentionality in design decision-making.

In the article "The Big Business of Unconscious Bias," Michelle Kim states that "the question isn't about whether it's effective . . . It's about how to make it effective." Olga Russakovsky in the article, "Dealing with Bias in Artificial Intelligence," states that "I do not think it's possible to have an unbiased human, so I don't see how we can build an unbiased A.I. system. But we can certainly do a lot better than we're doing." These two statements—making inclusion efforts in design effective and doing better—is feasible when sociocultural factors are taken into considerations for all designs generated.

Designers need to acknowledge that people are cultural beings; therefore, the practices of integrating cultural factors into design should be considered and emphasized.

Design is securely embedded in the culture of the users, and users are not just physical and biological beings, but also sociocultural beings.

1.5 NEW TECHNOLOGICAL PRODUCT DESIGN AND LAUNCH CHALLENGES

Product development and launch begins with the ability to bring together a talented team of professionals from various disciplines, such as management, engineering, accounting, and marketing. Together these disciplines can imagine, design, and place into the marketplace new products and technologies that can significantly improve the quality of life for society. Engineers as designers are extremely important to the developing and launching of new products, and therefore the training and education provided to engineers are of utmost importance.

The principal objective of engineering education is to prepare students to work as productive engineers in society. The learning experience focuses primarily on developing students' knowledge and experiences related to technical and analytical knowledge, skills, and abilities. Initiatives have emerged to develop a more holistic engineer that highlight an aspect of engineering education that is lacking; the development of essential nontechnical knowledge, skills, and abilities that can be useful in design and development of products to attract a more diverse customer base. However, changes are needed in developing engineering curriculum to develop the holistic engineer and designer for inclusivity. The article entitled "The People Part of Engineering" provides a well-structured framework that illuminates this challenge of equipping engineers to be able to address non-technical issues and concerns. This emphasis on objectivity in the fields/academic disciplines that drive technological design often highlight an appreciation of human diversity in industries, and the perspectives that the technologist brings to design activities. In creating products responsive to a diverse group of users, technology designers often do not have much other than themselves in mind or considerations for the limited population of familiarity.

As stated by Randy Reyes, Global Diversity and Inclusion Lead for Google,

> When people create products, they often bring the perspective of their own paradigm to the work, even with the best intentions for being inclusive. Therefore, when product managers are curious how the product they develop will be received by different groups, it is important to have those varied groups' perspectives be represented during the creation and testing processes.

BIBLIOGRAPHY

BBC News, Five Big Things That Have Made Apple, *BBC News*, August 2018. http://bbc.com/news/business-45044963

Bichard, J., Alwani, R., Raby, E., West, J., & Spencer J., *Creating an Inclusive Architectural Intervention as a Research Space to Explore Community Wellbeing*, Royal College of Art Helen Hamlyn Centre for Design, 2018

Brown, T., *Change by Design, How Design Thinking Transforms Organizations and Inspires Innovation*, Harper Business, 2019

Fila, N. D., Hess, J., Hira, A., Joslyn, C. H., Tolbert, D., & Hynes, M. M., The People Part of Engineering: Engineering For, With, and as People, *2014 IEEE Frontiers in Education Conference (FIE) Proceedings*, Madrid, Spain, pp. 1–9, 2014. doi: 10.1109/FIE.2014.7044106 https://accelerate.withgoogle.com/stories/meet-randy-reyes-global-diversity-inclusion-lead

Kim, M., The Big Business of Unconscious Bias, *New York Times*, November 2019. https://www.nytimes.com/2019/11/20/style/diversity-consultants.html

Langdon, P., Clarkson, J., Robinson, P., Lazar, J., Heylighten, A., & College, F., *Designing Inclusive Systems: Designing Inclusion for Real-World Applications*, Springer, 2012

Moalosi, R., Popovic V., & Hickling-Hudson, A., Culture-Driven Product Innovation, *International Design Conference—Design*, Dubrovnim—Cronatia, May 15–18, 2006

Perez, B., *The Hidden Value of Inclusive Design for Business and Innovation*, UX Collective, 2020

Persson, H., Åhman, H., Yngling, A. A., & Gulliksen, J., Universal Design, Inclusive Design, Accessible Design, Design for All: Different Concepts–One Goal? On The Concept of Accessibility–Historical, Methodological and Philosophical Aspects, *Universal Access in the Information Society*, 14(4), 2015, 505–526. http://dx.doi.org/10.1007/s10209-014-0358-z

Product Inclusion Leadership: Meet Salima Bhimani, Chief Equity and Inclusion Strategist, The Other Bets at Alphabet, accelerate.withgoogle.com

Rowe, P. G., *Design Thinking*, The Massachusetts Institute of Technology, 1987

Searle, M., & Executive Vice President of Arizona State, *Top Universities Join to Push Public Interest Technology*, Springer, 2019

Smith, C., Dealing With Bias in Artificial Intelligence, *New York Times*, January, 2020. https://www.nytimes.com/2019/11/19/technology/artificial-intelligence-bias.html

Swain, M., Kinnear, P., & Steinman, L., *Sociocultural Theory in Second Language Education: An Introduction through Narratives*, Channel View Publications, 2015

UK Design Council Website, *What is the Framework for Innovation? Design Council's Evolved Double Diamond*, UK Design Council Website, 2022 www.reference.com/world-view/examples-sociocultural-factors-16d3d81637f83a47

2 Humanistic Product and Technological Design
A Way Forward

2.1 INTRODUCTION

New product design is an important part of technological advances and is forever changing the way of life for consumers and continues to be a vital part of advancing the quality of life. Humans are the targeted consumers of the majority of products and technologies designed, developed, and sold within the global economy. Therefore, it makes sense that when designing products that are expected to be consumed or utilized by humans, special attention is paid to the product or technology to human interface. This type of interface can be viewed as useful, enhancing, or even detrimental if not designed appropriately. One must recognize that designing to encompass the diversity of users is not easy and requires a mindset that may not have been fully tapped into as of yet or that is difficult to invoke and sustain during idea generation and the design process. This is where humanist product and technology design come into play and should take root into the design process, starting with the research phase of the design.

Why begin with the research phase? Because it is within this phase where research or consideration is given to the potential users and questions are asked that will render the final product usable and safe for consumers. The ultimate goal should be to discover and address any humanistic impacts that may pertain to equal usability in the conceptual design phase, where intense collaboration and brainstorming takes place to provide a second look at the concept, design, and expected performance.

Humanistic design as discussed throughout this book refers to designing products to fit the needs of people. Explicitly, these designs are bringing together the human product interface to a point that the product provides value, improvement, and advancement to the human way of life. Humanistic design offers the potential users a level of assurance that the product will perform as intended, for the preponderance of users having minimized the possibility of "doing harm."

To illustrate some of the humanistic considerations that takes place in each of the four listed design phases, Figure 2.1 offers some considerations to help with the beginning of focusing one's thought.

Theorists and practitioners are starting to engage in frequent conversations about the necessity to ensure products, services, and technologies be designed with a diverse population usage in mind. Rob Girling and Emilia Palaveeva state the following:

DOI: 10.1201/9780367854720-3

Research Design Phase	Conceptual Design Phase	Design Development Phase	Design Complethion Phase
•Human needs that will be fullfilled •Population characteristics •Ccultural aspects	•Outline the process, product or technology •Include a diverse team to evaluate human needs discovered during the reasarch phase and any other intricacies that may be identified by team members •Do not approach this phase in a vacuum	•Humanistic aspects incorporated	•Humanistic aspects included and verified

FIGURE 2.1 Humanistic Design.

We need to be clear-eyed about what we are striving to do and minimize the chances of creating more problems than we are trying to solve . . . as a result, we will be poised to design systems that have minimum negative impact, create and sustain equity, and build on technological advances without disrupting the foundations of society.

And, becoming more "clear-eyed" requires greater focus and mindfulness to notions of exclusion that can often cloud technological design and development decision-making. Exclusionary practices, patterns, behaviors, and norms that are ingrained within the culture of engineering that, while unintentional, may lead to future technological solutions that do more harm than good and exclude segments of the population. By no means is it being suggested that what is being witnessed is deliberate. Nonetheless, these factors, individually or collectively, cannot be given a pass, with their consequences—*the disenfranchisement of segments of, humanity*—simply dismissed or overlooked. Table 2.1 provides several examples of ideas that prevent inclusive design.

Understanding and addressing questions of exclusion and equity within engineering cannot be more relevant in today's environment. However, as discussed by Erin A. Cech in her paper, "The Veiling of Queerness: Depoliticization and the Experiences of LGBT Engineers,"

there is a level of resistance within the profession of engineering to discuss and work to address issues of inequality in technology design and development.

Further, Cech states:

This resistance comes in part from the professional culture of engineering, the system of meanings, values, norms, and rituals built into and around engineering tasks and knowledge. One particular ideology within the culture of engineering, the ideology of depoliticization, misframes questions of inequality within engineering as marginal and largely irrelevant to "real" engineering work.

TABLE 2.1

Some Ideas that Prevent Inclusive Designs

Young adults are viewed as being smarter, more creative, energetic, and employable than older adults, children, and teenagers	Adults having college degrees are viewed as smarter than those adults not having college degrees.
Heterosexuals are more deserving of love, dignity, humanity, and companionship than people who identify as lesbians, gay, or bisexual.	English speaking people with dominant culture accents are viewed as being more intelligent than non-English speaking people.
People who are physically and mentally disabled are less intelligent than people whose abilities conform to the majority.	People who identify as Christians are more trustworthy and righteous than non-Christians.
Rich people are smarter, more trustworthy, responsible, and deserve aspiration and grace more that people who are poor.	Males are smarter, better leaders, stronger, more responsible, and honest than females.
Lighter skinned people are more deserving of love, affection, power, wealth, grace, and dignity than darker skinned people.	Cisgender people are more deserving of love, companionship, dignity, and humanity than people who identify as being transgender.

Thus, evolving within engineering design the requisite competencies to appreciate, understand, and respond to these issues of inequality in offering technological solutions that are beneficial to all is of importance. Humanistic Engineering could provide the necessary rigor within engineering design. In unpacking a more humanistic engineering approach within the engineering design space in grappling with these notions of inclusion, a framing for engagement emerges.

Engineers and designers are a product of their environment. In saying so, although well trained technically and competent to make design-related decisions, they bring to the workplace certain biases than can prevent them for thinking human centered and inclusive. Many of these professionals were exposed to some ideals such as the ones listed in Table 2.1. These ideals can have a significant impact on the thought process and designs produced by professionals. Humanistic design thinking seeks to properly catalogue and remove potential influence of these thoughts or ideals from the design decision making process so that each design represent exclusivity and equity for the preponderance of users.

2.2 HUMANISTIC ENGINEERING FRAMEWORKS FOR NEW PRODUCT DESIGN: BEYOND HUMAN-CENTERED DESIGN (HCD)

Traditional human systems integration (HSI) approaches to new product design often are not enough. "As technology becomes more ubiquitous, it is essential we consider the impacts on people, whether unintended consequences or designs that exclude certain groups or disadvantage them in some way," states Mark Searle, executive vice president of Arizona State in the *New York Times* article "Top Universities Join to Push 'Public Interest Technology.'" More inclusive and consequential approaches to new product design are needed. Technology is not neutral. In creating products responsive to a diversity of users, technology designers often do not have much other than themselves or others that identify with in mind. "Designers have been slow to respond to demographic shifts, designing for only a narrow segment of the population" and "the narrower those people's perspectives are—the more they design and code for people like themselves and shrug off any responsibility for the outcome." As detailed in a recent FastCompany article,

> If we want better tech products, we need to develop them with diversity in mind. This is one of the reasons why the market for diversity and inclusion technology—tools that prevent and addresses bias, harassment, and discrimination—is worth roughly $100 million. Consider, for example, that Apple's Health app initially left out a function that would have been helpful for approximately 50% of its users: tracking menstrual cycles.

Humanistic Engineering Frameworks for new product designs begins with a way of thinking that focuses on a couple of key inclusive design concepts that enhance the practicality of achieving a humanistic design to inform and produce designs that lead to products, services, and technologies that are useful and do not harm humans as a whole. A pictorial view of the key aspects of humanistic engineering is shown in Figure 2.4. These key aspects form the basis of inclusive design and are further discussed in the succeeding chapters. These tenets along with pertinent supporting information are listed in Table 2.2.

While nascent HSI umbrellaed more human-centered design (HCD) or design thinking framings to new product design such as inclusive design, humanity-centered design, or liberatory design thinking exists that can aid in better understanding and dealing with broader contextual considerations in design; their current bounds, in engagement, often constrain the technological design narrative, limiting design decision-making to the frames of present-day mental models. As highlighted by Sarmite Polakova and Audrey Cruchade, designers at Digital Agency Mirabeau in a Medium UX Planet piece,

> "The problem with the current Design Thinking practice is that it focuses on a small number of users and at times fictional personas. It asks fundamentally narrow questions, for example, "How might we rent out our homes?" But the question we forget to

TABLE 2.2
Humanistic Product Development Tenets

Tenet	Approach	Tenet Enablers	Tools
Equitable	Liberatory Design	Community-Based Participatory Design	EquityXDesign
Culturally Calibrated (**Contextual**)	Intersectional Design Culture-oriented Design	Reflexivity: Positionality Worksheet	A Designer's Critical Alphabet
Consequential	Pendulum Thinking	Systems Thinking Futures Thinking	Vision Concepts

ask should be: "How does this idea work for a white woman? Or for a Black gay man? For a child in a wheelchair? For a low-income family? Who else is going to be touched by this innovation?" or, often more importantly who will be excluded or further marginalized and/or disenfranchised by the given technology?

Thus, adequately grappling with diversity in technological design is truly complex and multilayered. More inclusive design practices are truly needed as we look to technological solutions that meets all of humanity's needs. Tools and techniques such as futures thinking offer a start that must be continuously explored and developed. There are gaps in offering more inclusive technological new product designs and consequences are being realized—*design and design practices have unintended consequences.* While these cannot be eliminated, what approaches like futures thinking offers are a means to

(a) Make visible biases

and

(b) Offer a methodological framework for challenging and addressing them

It is the intent that appropriately deployed and supported, futures thinking, and comparable approaches will foster more inclusive design thinking within new product designs that not only aids in both uncovering salient sociocultural considerations and offering a means by which to appropriately respond but, as such, provide design framings for technological product innovation.

2.2.1 LIBERATORY DESIGN APPROACH

In today's world engineers are challenged with demonstrating their innovative abilities through design of new products and services that place their corporations in a position of competitiveness as an industry leader. A comprehensive and promising approach to meeting this challenge is the use of design thinking. Design thinking is an approach to problem-solving that leads to the development of innovative products, technologies, and services. Liberatory design thinkers are empowered and feel free to utilize the design thinking approach to solve problems and explore new ideas that can lead to novel inventions in technology or services. These designers will often introduce ideas and designs that may be considered to some as nontraditional, different, or even far out from what is seen for their colleagues. However, with time the uniqueness of the discoveries of these designers will become the norm, and they will continue to push the envelope into a century of new designs. In many senses, these designers are considered pioneers of their time with an uncanny ability to change the quality of life for others.

This ability to explore and think outside of the box, as appropriate, can lead to many discoveries that can continue improve the quality of life for humanity. When engineers and designers are restricted to their myopic thought process due to unnecessary procedures and practices, they become limited in their ability to liberatory design products and technologies that are designed with the future and the human product interface in mind.

2.3 INCLUSIVE DESIGN

At the center of design products and technologies that will be useful and beneficial for the population of culturally diverse people are three key attributes. These attributes are shown in Figure 2.2. The three attributes are at the core of developing design that are humanistic and inclusive. The components of inclusive designs shown in Figure 2.2 when put into practice will assist engineers in designing products that can be used by a diverse population. These components must be in alignment and a cohesive part of the product design process to form an inclusive design. The three concepts represent an important trilogy in the inclusive design journey and process.

The inclusive design components will be discussed in more detail in Chapter 10 of this book.

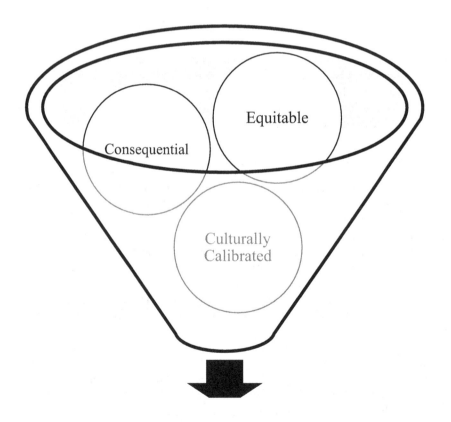

Inclusive Design

FIGURE 2.2 Inclusive Design Components.

2.4 HUMANITY CENTERED

Human centered design is a philosophy that places emphasis on a comprehensive approach to design with the expressed purpose of humanizing designs. Human centered design philosophy gives engineers a process and method to aid in solving complex design problems. Human centered design allows appropriate focusing on the human product interface which adds value to the resultant products from the consumer perspective. This type of focusing is not easy and for some may not seem natural and perhaps distracting as the designer must always be alert to, and focus on, two aspects (human usage and product performance) of design as opposed to one.

Current design practices in some instances seem to place more importance on product design and delivery and little to no attention to how the products will impact the various consumers that may purchase and attempt to utilize the product. This is obvious because of the many cases revealed over time that detailed the failure of products as it relates to consumer usage. These failures have created adverse effects to users from little harm to catastrophic harms, such as physical or mental injuries. A great example of a failed design is a simple step ladder. Although many step ladders are designed to be functional, the simple design addition of a top and side handrails changed the level of safety provided to the consumer. Employing designs that takes into account the product human interface is destined to reduce or eliminate many of the adverse impact experienced thy users.

2.5 FUTURISTIC DESIGN THINKING: A PATHWAY FORWARD

The thought and concept of designing for the future is gaining traction across the globe; however, many struggle with how to design for the future. Design thinking is traditionally focused on identifying issues or requirements using a human-centered approach, whereas futuristic methods seek to foresee the future from a longer-term perspective. Future thinking has been defined as an iterative process that helps an individual consider a wide range of possibilities and potential outcomes. A more formal definition of futurology is "the scientific study of possible probable, and desirable future developments, the options for shaping them, and their roots in past and present." The pathway forward to engaging in the development of futuristic design involves a comprehensive approach that makes use of a thinking process that involves strategic and intentional actions of the designer put into motion for each design that they embark upon. Several of these actions are shown below in Figure 2.3 and discussed further in this section.

Each of the characteristics of futuristic thinkers is critical to achieving the process in totality. These characteristics are exercised throughout the process and are not listed in order of engagement. As such, a futuristic thinker is skilled in knowing when to invoke these characteristics.

> *Open to possibilities*—Being open to possibilities allows the designer or engineer to feel free to explore various ways of discovering and developing new products, services, or technologies that create value for their customers. Exploration of different avenues are welcomed even if the idea is presented from others. Being open to possibilities provides the steppingstone to readily accept and

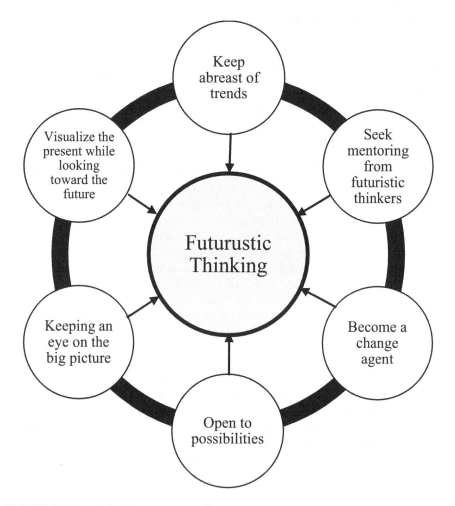

FIGURE 2.3 Futuristic Thinkers Evolution Paradigm.

adapt to changes. There are several elements that should be focused on and held close in aiding and keeping an open mind to embrace different aspects of design and possibilities for the future. Several of these elements are listed along with the potential benefits of invoking each element.

Presented in a cyclic format, Figure 2.4 serves as a reminder that these elements are connected and are continuously in motion to achieve the desired openness and freedom to explore, discover, and create. More elements may be added; however, the significant ones are referenced in Figure 2.4 and briefly highlighted in more detail below.

- Replace negative thoughts with positive thoughts
  - Reduces stress and provides the drive to approach the future
- Challenges traditional beliefs and ideas
  - Reduces possibility of myopic vision

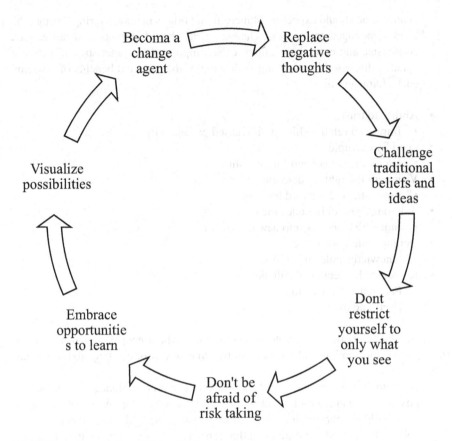

FIGURE 2.4 Open to Possibilities Cycle.

- Do not restrict yourself to only what you can see
  - Focuses on what you can only see is limiting progress
- Do not be afraid to become a risk taker
  - Leads to better designs and products
- Always embrace the opportunity to learn and continue to learn
  - Knowledge leads to creative discovery
- Visualize the possibilities
  - Facilitates hope and direction
- Become a known change agent
  - Influences the trend present and future

Become a change agent—Futuristic thinkers by default are change agents because of their ability to adapt, design novel products and technology, and push the envelope of traditional design thinking. A change agent is someone who is skillful in initiating and/or managing change in a group or an organization. Change agents are often considered visionaries as they continuously seek ways of doing things differently and more efficiently. Change is constant; therefore, there should be an expectation that change will occur, and when dealing with change, one should expect resistance. To aid others in maneuvering through the change management process, change agents should embrace resistance, seek consensus, and exert their influence. Six important characteristics of a change agent is discussed below along with some of the expected benefits of invoking each characteristic.

- Ability to trust
  - Improved relationship and increased productivity
- Lead by example
  - Facilitate respect and followership
- Visionary thoughts and actions
  - Futuristic and forward looking
- Ask tough provoking questions
  - Inquisitive and open to new discoveries
- Demonstrate competence
  - Knowledgeable and skilled
- Respected by peers and colleges
  - Facilitate followership
  - Facilitate trust

These characteristics can create an environment where people are eager to learn, free to share information, and open themselves to new way of thinking and new ideas.

Seek mentoring for futuristic thinkers—It is wise to seek guidance from those who possess the characteristics that you are seeking to develop. A mentor can assist with building current skills or learning new skills, gain confidence in one's ability, and provide an individual that can serve as a sounding board for various issues, decisions, and thoughts. A futuristic mentor can provide assistance in focusing the mentee in areas of thinking and provide suggestions on the path to moving forward in cataloging and prioritizing their thoughts can be useful.

- Exposure to new ways of thinking and ideas
  - New possibilities and discoveries
- Develop futuristic thinking skills
  - Enhance strategic capability
- Obtain advice on developing strengths
  - Capitalize on your successes and failures
- Obtain advice on overcoming weaknesses
  - Turning weaknesses into strengths
- Build visionary skills
  - Ability to anticipate future potential issues and devise a plan to circumvent
- Embrace the possibilities of tomorrow
  - Change is a constant event

Keep abreast of trends—Understanding and keeping abreast of product and technological trends can reveal market opportunities and help maintain sustainability of current product lines. It also can provide insight into the challenges a particular segment of the population may have in adapting to or use of various products and technologies. Losing touch with technological trends can ultimately lead to extinction of a company due to loss of customers and products that formerly were considered valuable.

- Keep eyes and ears open
  - Provides insight into the future on what may come
- Social networking
  - Collaboration and new business possibilities
- Keep abreast of evolution and trends in technology and products
  - Gives insight on market path and longevity
- Constantly read—trade journals, articles, etc.
  - Awareness and knowledge are power
- Observe people who are trend setters
  - Provides clues on future activities

Visualize the present while looking toward the future—Formulation of a mental image of how a design will work in the future or impact future users. The imagination plays a significant role in a person's ability to creatively formulate thoughts that go into the design process. Having the ability to visualize the present invokes three key aspects or characteristics that must be exercised: develop mental images, intellectual insight, and foresight ability.

- Develop mental images
  - Visualizes the possibilities
- Intellectual insight
  - Allows inward looking that can help identify biases
- Foresight ability
  - Seeing beyond what is in front of you provides opportunities to chart future course

Keeping an eye on the big picture—It is important not to become myopic and paralyzed by the present. Allowing oneself to become myopic has a significant impact on the ability to exercise imagination, intellectual insight, and foresight into what is happening in the present and may happen in the future. Taking the time to ensure an understanding of the concept or situation in detail is a signification part of not allowing the big picture to be lost on distractions or unimportant attributes or situations.

- Avoid becoming myopic
  - Avoid missing what is happening in the present and impact on the future
- Focus on the vision
  - Allows broadening of views
- Avoid distractions
  - Keep focus on the vision

2.5.1 Design Thinking Empowered by Futures Thinking

Designing better products, technologies, and services is the optimal goal of invoking design thinking. This concept of design thinking is being adopted by large corporations to facilitate product quality and sustainability. The concept of design thinking has also been incorporated into engineering processes and into business management practices. Design thinking becomes easier when the need of the customer is known and understood. This means that design thinking not only places focus on the product that will be produced, but also the impact on its human users. As alluded to many times, design thinkers skillfully use cognitive, strategic, and pragmatic abilities to ensure that designs are practical for users.

Being futuristic really means having the right mindset coupled with knowledge and practice. The mindset of a futurist is

- Open and embraces discipline of actions
- Always considers the long-term consequences of their actions
- Demonstrates a commitment to all human endeavors

Future thinking requires a person to invoke the mindset toward forward looking applications using strategic and systematic approaches. When design thinking is integrated and empowered by futures thinking a solution oriented, human centered, and inclusive professional is born, and the products produced have the capability to serve the needs of a diverse population.

2.6 APPLIED LEARNING

2.6.1 Case Study

Read the following case study and answer the following question.

Company XXLX designed a product that was intended to be used to measure the heart rate of consumers. The product has a bulky exterior and weigh about 25 lbs. Shortly after product launch, the company became aware of feedback that the product casing was too bulky and not feasible for some segment of the population to move the unit from place to place. For example, senior citizens had difficulty in using the product due to difficulty experienced in handling the unit. The product design team consisted of white males between the ages of 20–25 with less than three years in the work environment. Two of the team members lived with grandparents who were not tolerant of people who were different, such as skin tone or language. One outspoken team member communicated that they believed the design should accommodate people who were young. presumably healthy and athletic just like the members of the team.

Question: How might the team members and design team make-up impede success of developing a product that meets the needs of a diverse population?

Additional questions:

1. List the tenets of humanistic product development.
2. Describe the humanistic framework for new product design.
3. What are the components of inclusive design and how does each component enhance design thinking and practices?
4. List at least four elements of futuristic thinkers. Discuss the use of each element in the new product design process.
5. Analyze Figure 2.4 and explain the importance of each element and why it is necessary to think of each element in terms of being cyclical.
6. List the characteristics of a change agent. Discuss how each characteristic is key in the change management process
7. What is the role of a mentor when developing futuristic thinkers?

BIBLIOGRAPHY

Cech, E. A. The veiling of queerness: Depoliticization and the experiences of LGBT engineers. *Paper presented at 2013 ASEE Annual Conference & Exposition*, Atlanta, June 2013

Cecily, S. *Think Like a Futurist: Know What Changes, What Does not, and What's Next*, John Wiley & Sons Incorporated, 2012

Garreta-Domingo Muriel, S. P. B., & Hernández-Leo, D. Human-centered design to empower "teachers as designers." *British Journal of Educational Technology*, Vol. 49, No. 6, 2018, pp. 1113–1130. doi: 10.1111/bjet.12682

Hehn, J., Mendez, D., Uebernickel, F., Brenner, W., & Broy, M. On integrating design thinking for human-centered requirements engineering. *IEEE Software*, Vol. 37, No. 2, 2020, pp. 25–31. doi: 10.1109/MS.2019.2957715

https://medium.com/equity-design/the-big-10-1ideas-that-fuel-oppression-97d7200929f9

https://uxplanet.org/can-speculative-design-make-ux-better-design-trend-4-fue;-oppression--4-ce8d13148e5d

https://uxplanet.org/can-speculative-design-make-ux-better-design-trend-4-4-ce8d13148e5d

Srinivasaraghavan, N., Gurusamy, K., & Keighran, H. Forward Thinking: An Integrated Framework for Formulating Vision, Strategy and Implementation. In: Marcus, A. (eds) *Design, User Experience, and Usability: Design Thinking and Methods. DUXU 2016. Lecture Notes in Computer Science*, Vol. 9746, Springer, 2016. https://doi-org.ezproxy.proxy.library.oregonstate.edu/10.1007/978-3-319-40409-7_1

Tauke, B., Smith, K., & Davis, C. (eds.). *Diversity and Design: Understanding Hidden Consequences* Routledge, 2015

Wachter-Boettcher, S. *Technically Wrong Sexist Apps, Biased Algorithms, and Other threats of Toxic Tech*, W.W. Norton and Company, 2017

Winchester, W. Inclusive and Consequential by Design: "Futurefying" New Product Development (NPD) Through Vision Concepting. *Insight International Council on Systems Engineering*, Vol. 22, No. 3, 2019–2010, pp. 49–51

www.fastcompany.com/90149212/beyond-the-cult-of-human-centered-design

www.fastcompany.com/90532486/im-a-latina-veteran-wo-works-at-amazon-diversity-isnt-about-checking-a-box

www.nytimes.com/2019/03/11/technology/universities-public-interest-technology.html

3 Speculative Design in New Product Design

Enacting Humanistic Engineering Approaches

3.1 INTRODUCTION

Product and technology development are constantly evolving and increasing in importance in improving the quality of life for consumers near and far. Success in launching new products requires engagement in innovative thoughts, ideas, and actions. Innovative ideas and solutions are key for the sustainment of an everchanging and prosperous world culture. Innovation involves changes that enhances technologies, products, and services that produce positive results for customers. To achieve success, product innovation requires the support of a team of professionals with a diversity of skills, knowledge, and intuition. According to Rainey 2005, the key aspects or strength of product innovation includes "(1) Examining the needs for new products, processes, and services. (2) Determining the proper direction and fit for new products. (3) Establishing the appropriate game plan of the entire management system for developing and commercializing new products. (4) Selecting new-product opportunities for investment. (5) Enhancing the organizational capabilities to create successful new products. (6) Creating the new product and executing the new-product design) program." Successful development and launching of new products require a knowledgeable team that is dedicated and focused on product innovation.

New product development is not easy as it requires many activities that will be performed by a diverse discipline of professionals. These professionals will be engaged in, for example,

- The generation of new ideas
- Gaining knowledge and foresight into market trends
- Determining the customer needs
- Evaluating strengths and weakness of products and technologies currently being utilized

3.2 TEAMING IN NEW PRODUCT DEVELOPMENT

Teaming is a concept that has been embraced across the globe by leaders, theorists, and practitioners because of the recognition of the benefits that are gained through cross-functional collaboration. A team is viewed as a group of people

DOI: 10.1201/9780367854720-4

with the required complement of skills working in concert to complete a task or a project. Each team member is accountable for the team performance thereby increasing each member's engagement. New product development does not happen in a vacuum nor is it the work of one individual. Developing new products takes a team of highly knowledgeable and skilled engineers and professionals to deliver a product to the market that will serve the needs of customers and enhance quality of life. The team will consist of professionals from disciplines such as engineering, finance, project management, project controls, administrative, and more.

The benefits of embracing teams in new product design include

- Diversity of thoughts and experience
- Enhancement of developing products that are inclusive and valuable to a diverse population of users
- Opportunity for team members to learn from each other
- Increase trust in organizations
- Enhance defensible and sound decision-making

3.3 HUMANISTIC ENGINEERING

When referring to humanistic engineering, focus is being made on engineering practices that take into account the people aspects of engineering utilizing knowledge gained from the field of social sciences and humanities. Humanistic engineering is being described as the integration of various sciences and disciplines such as humanities and social science coupled with engineering. The people aspects of engineering are not always at the forefront of thought when ideas are formulated and explored and when designs are contemplated and is oftentimes considered more completely at other phases of the new product development process. Figure 3.3 provides a pictorial view of the complexity of deploying humanistic engineering. Many of the complexities come into play as a result of potential burden of deploying design while using knowledge in the disciplines represented in Figure 3.4. On the surface developing these designs can seem daunting, especially when engineers are not traditionally trained in other areas beyond engineering.

Table 3.3 lists the benefits of the three disciplines working together hand in hand to design the optimal products for current and future usage. The social science discipline can include disciplines such as psychology, sociology, economics, communication, philosophy, and ethics. The use of humanistic engineering has widespread implications and benefits for new product design Although incorporating the social sciences and humanities discipline is not easy for most engineers, as many of them have not been trained to focus on what we will refer to as the "all in all people aspects."

Engineers can be trained to skillfully combine the three disciplines to imagine and design new products that can have an enormous impact on improving the quality of life. A review of Figure 3.1 and Table 3.1 demonstrates many of the benefits of employing humanistic engineering in new product design.

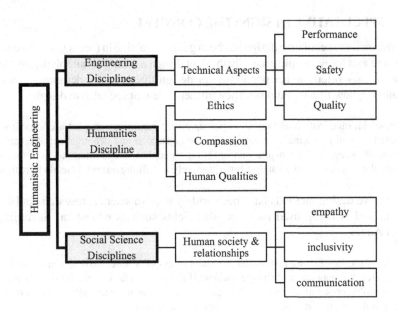

FIGURE 3.1 Humanistic Engineering Paradigm.

TABLE 3.1
Humanistic Component Design Implications

Engineering	Humanities	Social Sciences
Defined: "Engineering is the use of scientific principles to design and build machines, structures, and other items, including bridges, tunnels, roads, vehicles, and buildings. The discipline of engineering encompasses a broad range of more specialized fields of engineering, each with a more specific emphasis on particular areas of applied mathematics, applied science, and types of application. See glossary of engineering" (Wikipedia)	**Defined:** "Humanities are academic disciplines that study aspects of human society and culture" (Wikipedia).	**Defined**: "Social science is the branch of science devoted to the study of societies and the relationships among individuals within those societies" (Wikipedia).
Significance: engineers provide the ideas and conceptual design that led to new product development using their technical knowledge gained through schooling and experience. As a part of the technical aspects of new product design, the engineer must ensure that the product performs as design is of quality to support customer needs and can be used without inflicting harm to users.	**Significance**: having knowledge of the various aspects of the societies in which humans reside and the culture that they reside in is important in design considerations that should be flushed out and explored beginning with the idea phase. Understanding the diversity that people bring to the table can be instrumental in the appropriateness and inclusivity of design and use.	**Significance**: knowledge in various aspects of social sciences can help engineers design products that can enhance relationships social interactions and workplace efficiency.

3.4   SPECULATIVE DESIGN: THE CONCEPT

Discussion into speculative design has been gaining traction in recent years. According to Dunne and Raby, people generally think of design as being about solving problems; however, they offer another possibility for design that is to use designs as a way of speculating how things could be. They further state that speculative design

> thrives on imagination and aims to open up new perspectives on what are sometimes called wicked problems, to create spaces for discussion and debate about alternative ways of being, and to inspire and encourage people's imaginations to flow freely. Speculations can act as a catalyst for collectively redefining our relationship to reality.

Speculative design methods have been widely used in science, research, and development and is being used more in other fields such as education. According to James Auger,

> "one of the key factors responsible for the success of a speculative design project is the careful management of the speculation; if it strays too far into the future to present implausible concepts or alien technological habitats, the audience will not relate to the proposal resulting in a lack of engagement or connection.

Speculation in design in brief is a theory-based practice where conjecture, abstract reasoning, and intellectual speculation is used to develop ideas into designs that advance technology and products while shaping the future for consumers. The characteristics that are at work within each of the design components are shown in Figure 3.2. There remains the challenge of not going too far into the future to where the ideas and use of speculation seems unrealistic or unachievable.

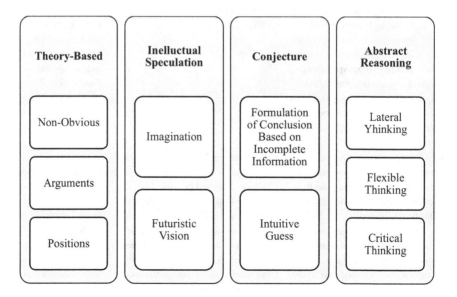

FIGURE 3.2 Speculative Design Components.

The four elements of speculative design presented in Figure 3.4 are the underpinning mechanisms that place the practice in motion and keep the practice ongoing to achieve the products that are revilement in the future.

3.5 ENGAGING SPECULATION IN DESIGN

Considering the major components of speculation in design and their role in engaging speculation in design, a clear understanding of these four components is crucial.

3.5.1 THEORY-BASED

Decisions or assumptions made based on theory is lacking experimental data. The other three components of speculative design are considered as theory-based and can be used as predictors or complement decisions that have not been subjected to testing or experimentation. According to Hannay et al. empirically based theories are often viewed as foundational to science and many arguments exist that favor the use of theories. Some reasons theories are accepted and used include the following:

* Theories present a conceptual framework that permit knowledge structuring in a succinct manner.
* Theories facilitate communication or ideas and knowledge.
* Theories offer a level of abstraction that allow the independent generalization of knowledge without considerations for time and place.
* Theories allows analytical generalization.

Empirically-based theory is usually based on observation and experience and as such it can be instrumental when used in the speculative design process.

3.5.2 INTELLECTUAL SPECULATION

An intellectual person engages in critical thinking, relevant information gathering, and the appropriate level of reflection to make decisions that are appropriate and defensible. Speculations that are made based on these attributes are absent emotions and are generally deemed credible. Intellectual speculations have the capability to enhance the design process based on the designer or engineer using their intellect to drive thoughtful ideas that lead to novel designs that will produce novel products. Recognizing that not all designers, engineers, and leaders will accept the role of intellectual speculation as a component in design, there must be thoughtfulness in formulating and flushing out speculations that are fueled or driven by an individual's intellect. In addition, these designers and engineers must be respected and have credibility before pitching their ideas and designs.

3.5.3 CONJECTURE

Conclusions drawn and decisions made as a result of conjecture are common and visible in many disciplines that include science, math, and engineering. Deamer defines conjecture as an opinion or conclusion that is formed on the basis of incomplete information and involves the imagination as well as requires some level of creative effort. As such, there is no surprise that the use of conjectures in speculative design will add value to the process and enhance idea development and creativity in pushing the envelope in achieving products that may not be otherwise developed, marketed, and produced. Conjecture when used during speculative design, allows the designer while formulating their ideas leading to a design decision to request that certain information be considered as truth based upon previous or current related facts.

3.5.4 ABSTRACT REASONING

Abstract reasoning is defined by G., Fuhrmann, D., Knoll, L. J., Pi-Sunyer, B. P., Sakhardande, A. L., & Blakemore, S. J., "as the ability to solve novel problems without task-specific knowledge and a core mechanism of human learning." Khorshidi, M., Shah, J. J., and Woodward, J. defined abstract reasoning as the "ability to quickly identify relationships patterns and trends, integrate this information, and apply it to solve problems." This reasoning technique involves formulating beliefs concerning the nature of ideas, material, and processes. Utilizing abstract reasoning is appropriate for tasks that have little data that is available and for problems that do not require complete accuracy in results. These two scenarios are often in play during the design process and are commonly used during the development of design ideas and problem-solving. In addition, during the early stages of design, there are times where little information is known that is considered precise or concrete. These authors defined abstract reasoning and recognized the problem-solving aspects associated with abstract reasoning and its importance in the area of design. This importance holds true in the application of speculative design.

Abstract reasoning plays an important role in speculative design for reasons such as

- It enables the designer and engineer to think and develop ideas that are beyond what is obvious.
- Enables designers and engineers to imagine novel ideas and solutions to approach development of designs that may not necessarily be envisioned
- Assists designers and engineers in understanding and developing theories
- Facilitates divergent thinking that enables generating creative ideas

3.6 SPECULATION IN DESIGN: THE OPPORTUNITIES AND CHALLENGES

Using speculative design methods has many challenges; however, the opportunities associated with method usage perhaps outweighs any challenges. The challenges involved can be managed and overcome with a strategy, patience, and persistence of the designer. **One of the most import aspects that may determine the success of speculative design projects is the ability to strategically manage one's speculation.** However, if the design is viewed as moving too far into the future by presenting concepts that are implausible, the audience will be lost and will not be able to relate

TABLE 3.2

Speculative Design Challenges & Opportunities

Speculation in Design Opportunities	Speculation in Design Challenges
Provides an opportunity for designers and engineers to utilize innovation and creativity in new product development (pushing the envelope)	Some design may appear to be too futuristic and a segment of the population as Sci fi. As such, the designer and the design may not be valued, and there may be some difficulty in selling or gaining support for the design and the resultant product.
Takes design and product development to a new level	Training engineers and designers to accept the practice of speculation in design and utilize the concept in their decision-making process
Presents an alternative approach to traditional design thinking that can be used across the globe to enhance product development and performance	Some designs may seem unrealistic and difficult to accept by consumers.
Anticipating and reactively responding to future needs	Some segment of the population will not accept or agree with the need for the future design.

and connect to the concept. Table 3.2 lists some of the challenges and opportunities of engaging speculation in design.

3.7 NARRATIVE-DRIVEN INNOVATION

Narratives used in innovation can help connect the customer with the history behind the product and the vision of the designers and engineers. As such, it is incumbent of leaders and design engineers working together to focus on and develop narratives that are in alignment with the technology and product early in the design phase. An important attribute of developing narratives that align together cohesively will tell a complete and succinct story of the history of the innovative process that led to the end product. **A well-designed narrative can help connect customers with the product, and this connection can increase end product support and product usage**. During the design process, storytelling can be used to understand the problems that consumers face when using a product or a service. This knowledge is then used to formulate a solution that is then tested by the consumers. Storytelling during the design process can be implemented using techniques and methods, such as digital and visual storytelling, scenario generation, story boarding, and animation. Storytelling is a powerful communication tool that enables a better understanding about people's emotional experiences.

Storytelling can be one of the most effective forms of communication that we have at our disposal and is important in persuading others to accept and support our ideas. According to Biesenbach, a good story has three benefits. The three benefits of storytelling are

1. **Help win hearts**—capture attention, inspire. and motivate the listener
2. **Change minds**—convince others to retreat from their ways of thinking and accept the thinking of the storyteller
3. **Get results**—persuade others to act in a manner that will aid in goal achievement

Storytelling can help gain support for and connect others with a design especially when the design is speculative and futuristic.

3.8   VALUES IN DESIGN: DIVERSIFYING DESIGN IMAGINATIONS

Diversity in design increases the feasibility of wide-scale usage. When imagining new designs, aspects of that product must be subject to the imagination to include the diverse population that will be served by the product. During a talk given by Sengers Pheobe entitled, "Diversifying Design Imagination" on June 13, 2018, she stated that

> the act of designing technologies does not simply create functionality; it also offers possibilities for action, ways of looking at the world, and modes through which we can relate to one another. How we design technologies reflect what we value; who we think is important, and in what ways; which places, people and possibilities are in our imaginations, and which are not.

Sengers' position seems to believe the importance of ensuring diversification of design starts at the imagination phase to ensure functionality of technology for different groups of people. To achieve this, there is a requirement that the designer open their minds to the needs of people that perhaps they may not identify with or understand their lifestyle or history. She further pointed out that

> technology design is dominated by a narrow demographic: predominately white and Asian, white collar, highly educated, urban. These designers' ways of imagining new technological worlds are shaped by the worlds they themselves know and value, which are only a small slice of global ways of being.

She further states that

> as technology design is being increasingly engaged in around the world by and for people outside this demographic, local adaptations are frequently judged and limited by what makes sense from the perspective of Silicon Valley and other urban high-tech centers. Supporting the rich diversity of human experience requires explicitly identifying and appreciating values and experiences outside of mainstream technology design logics.

There are many values gained through ensuring that diversity is included in the design process, beginning with the stage where the technology or product is first imagined. These values are beginning to become the focus of many leaders and engineers across the globe. These values include

- Yielding that a product can be enjoyably used by a diverse population
- Opening the eyes of designers and engineers to the importance of embracing diversity
- Producing better and safer products
- Increasing profits for companies
- Increasing or improving a company's reputation as one of inclusion
- Increasing trust for products developed by the company by customers

3.9 NEW PRODUCT DESIGN AND BUSINESS ORGANIZATIONAL STRUCTURES

The structure of the business organization that is responsible for new product design is useful in providing formality and consistency in implementing programs and tasks that are key in enabling achievement of organizations and project goals which lead to a successful product launch. Effective business organizational structures are dependent upon the objective and strategy of the company. **The organizational structure can help set the stage for new product design and development activities by facilitating culture, alignment of communication channels, defining roles and responsibilities, determining how effective and efficient the organization functions, outlining, and communicating decision-making channels, communicating authority of the leadership team, and the quality of services provided to customers.** Designing the appropriate business organizational structure is important to ensure that the goals of the organization are met. When the focus is on product or technology development, some organizational business structures can enhance efficiency and goal achievement better than others. **The structure of an organization is key in defining how activities are aligned and the methods used by management to control business functions and workers to achieve the goals of the organization.** However, no matter what business organizational structure is put into place, any structure can work as long as the members of that organization have a desire for the structure to work and put forth the effort to ensure success. Utilizing an organizational system that is not complementary to the mission can be costly in terms or productivity, resource needs, communication, product development, launch failure, and cost. When focusing on new product design, there is one business organizational structure that may provide the best opportunity for ease of success. This structure is known as a divisional organizational structure. In a divisional system, all of the disciplines, including designers and engineers, are completely dedicated to a product line and provide their undivided attention to ensuring success of that product. A traditional divisional organizational system is shown in Figure 3.3.

Although divisional structures are efficient when it comes to product focus lines, it can be viewed as costly because of duplication of functions across a company due to dedication of support for each product line. This duplication of resources is

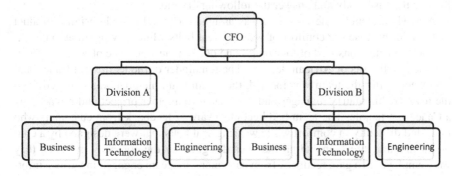

FIGURE 3.3 Divisional Organizational Structure Example.

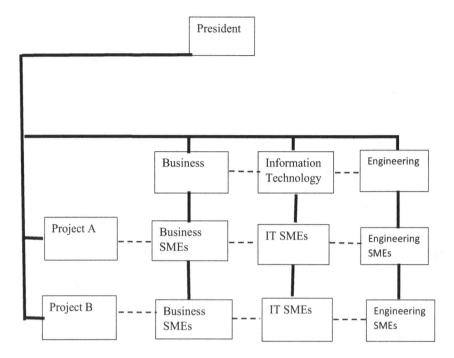

FIGURE 3.4 Matrixed Business Organizational Structure Example.

viewed as a major drawback of using this system, and some may be tempted to use an organizational matrix system to provide project support. In a matrixed business organizational system (Figure 3.4) resources are shared, and employees can support several projects, making it difficult to completely focus on one project or commit to a schedule that can support all tasks efficiently.

Additional discussion on organizational systems can be found in Chapter 6.

3.10 APPLIED LEARNING—CASE STUDY

Review the case study and answer the following questions.

A newly assembled design team was assigned to develop and design a product that would increase the comfort of wearing seat belts while driving an automobile. The design team consisted of four males and one female, one male of Asian descent, and the female was of African descent. The remainder of the team were Caucasian. Two weeks after the team was formed, the Asian team member was removed from the team by his matrix manager and reassigned to another project and replaced by a Caucasian female. The team had become aware of feedback from customers who indicated difficulty in comfort while wearing seat belts that were designed by a couple of manufacturers, including the ones designed by the company in which they were employed. To gain feedback from customers, the team engaged in various information gathering techniques including storytelling.

The team worked together to design a solution that was presented to the leadership team. The leadership team approved the design, and the new product was developed, mass produced, and sent to the automobile makers. Within six months, complaints started coming in indicating that the new seat belts caused a rash on the neck area as the belts do not fit the posture of many users as they are designed as "one size fits all" with minimal opportunity for adjustments.

1. What are the benefits of storytelling?
   a. How could storytelling be effectively used to gather information that can be used to improve the product?
2. Discuss the makeup of the design team and how the makeup might enable or hinder optimal success in developing an idea that can be beneficial to a large diverse population.
3. How can speculative design concept be used in solving the design problem that will render wearing a seat belt comfortable and safe for users?
   a. What must be avoided when imagining and introducing a new design?
4. How would diversifying design imagination ensure design and development of a suitable product?
5. In the scenario, one team member was replaced by their matrix boss.
   a. Describe the benefits and impact of the matrix organizational system on new product development.
   b. How would a divisional system have potentially impacted new product development in the scenario?
6. How can or should humanistic engineering be used by the team in development of a new product that enhances safety and provides comfort?
7. Define and discuss the speculative design concept and its impact on new product development and design.

BIBLIOGRAPHY

Anthony, D., & Fiona, R. *Speculative Everything: Design, Fiction, and Social Dreaming*, The MIT Press, 2013

Chierchia, G., Fuhrmann, D., Knoll, L. J., Pi-Sunyer, B. P., Sakhardande, A. L., & Blakemore, S. J. The matrix reasoning item bank (MaRs-IB): novel, open-access abstract reasoning items for adolescents and adults. *Royal Society Open Science*, 6(10), 2019, 190232. https://doi.org/10.1098/rsos.190232

David, L. R. *Product Innovation Leading Change Through Integrated Product Development*, Cambridge University Press, 2005

Deamer, D. Conjecture and hypothesis: the importance of reality checks. *Bullstein Journal of Organic Chemistry*, 13, 2017, 620–624. https://doi.org/10.3762/bjoc.13.60

Hannay, J. E., Dag I., Sjoberg, K., & Dyba, T. A systematic review of theory use in software engineering experiments. *IEEE Transactions on Software Engineering*, 33(2), 2007, 87. doi: http://dx.doi.org.ezproxy.proxy.library.oregonstate.edu/10.1109/TSE.2007.12 https://dl.acm.org/doi/pdf/10.1145/3196709.3196823

Hynes, M., & Swenson J. The humanistic side of engineering: considering social science and humanities dimensions of engineering in education and research. *Journal of Pre-College Engineering Education Research (J-PEER)*, 3€, 2007, Article 4

James, A. Digital creativity: crafting the speculation. *Digital Creativity: Design Fictions*, 24(1), 2013, pp. 11–35

Khorshidi, M., Shah, J. J., & Woodward, J. Applied tests of design skills—part III: abstract reasoning. *Journal of Mechanical Design, Transactions of the ASME*, 136(10), 2014, 101101. https://doi.org/10.1115/1.4027986

Rob, B. *Unleash the Power of Story Telling: Win Hearts, Change Minds, Get Results, East-lawn Media*, Orrington Ave, 2018 www.designorate.com/the-role-of-storytelling-in-the-design-process/

4 Design Lens and Vision Concepting
Advancing Speculation in Design

4.1 DESIGN IMAGINATION: FROM IMAGINATION TO IDEATE

Imagination is an important aspect of human thinking and is just as important to design. When we allow ourselves to imagine, we gain a visual abstract that can take us into a new direction where we can investigate new possibilities. A venture into the unknown is initiated by and begins with the imagination. An integral part of human thinking is the ability to develop a visual mental model that can be explored and expanded. The exploration of these models can be flushed out through team discussion and brainstorming sessions. **As depicted in Figure 4.1, the imagination sparks thoughts that help target the flow of ideas that can be visualized as options providing foresight that facilitate wisdom of what can be in the future.**

Folkmann states in his article "Unknown Positions of Imagination in Design," 2014 that "imagination deals not only with making and materializing new entities of meaning—for example, in a product or a design solution—but also with reflecting and integrating elements of unknown." A point to note is that during the vision stage of the ideate process, a collage of activities is happening to ensure reality of the vision is realized as an output to the ideate process. However, as will be discussed, the implications of our proposed engagement of futures thinking informs, ultimately, both early-stage design research and evaluative methodologies.

According to Stolterman and Nelson 2000, "in order to create, one must have the ability to imagine. Imagination is required within all fields of design."

4.2 DESIGN LENSES

To successfully produce designs that are capable of obtaining futuristic value and applicability, serve the needs of a diverse population, and increase value for consumers, another valuable element must be added to the design process. This valuable element is called design lens and can be applied at each design stage, but is most beneficial at the idea development stage of the design process. Design lens places the team in the environment in which their product will be used and assists them in opening their minds to consider people from the diverse that will be touch by their product. **Design lens is defined as a set of conditions, thoughts, actions, or activity that a designer or engineer considers that can influence their perception, comprehensions, and evaluation of concepts or activities.**

DOI: 10.1201/9780367854720-5

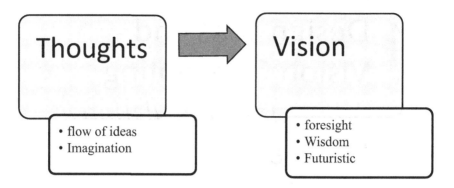

FIGURE 4.1 Thoughts to Vision (Ideate).

Design lenses offer a perspective to facilitate and enhance an inclusive view of a given product, process, or design dilemma. In operationalizing the concept of a design lens, we use the analogy of glasses. Design lenses when used to its fullest can provide a means by which to engage various dimensions of inclusion (e.g., gender, culture, race). The lens enables the product designer or engineer to consider the design activity and solution from different perspectives. Design lens makes clear and transparent those concepts, perceptions, and practices that could produce more inclusive decision-making and yields designs and products that are more representative of an inclusive population.

Design lens can equip the designer or engineer with the ability to view through the eyes of the end users; thereby, allowing the engagement of the appropriate filter (e.g., age, culture), bridging gaps between the designer's or problem solver's perception and the end user's perception, making visible those things (e.g., biases, stereotypes, and/or assumptions) that may be preventing a more inclusive design. Because a design lens only allows the view of a user experience through the eyes of one principle at a time, multiple lens will most likely be necessary for usage in order to design, develop, and deploy a truly inclusive product. Design lens is a key component of reducing myopic viewing during design, solution seeking, and the decision-making process.

Each design lens has the capability to challenge the designer or engineer to consider the design and problem-solving activities from a different perspective. This will make transparent those stereotypes or preconceived notions that can underpin decision-making throughout the design lifecycle and problem-solving process, from problem identification and exploration to solution implementation. Additionally, **selection and deployment of the appropriate design lens equips the designer with the ability to see through the end user's eyes, providing an opportunity to bridge gaps between the designer's or problem solver's perception and the end user's perception, making visible those things that may be clouding an equitable response**. Table 4.1 lists examples of different types of design lens.

4.2.1 When to Engage Design Lens?

Design lenses can, and should be, actively engaged during all phases of product development. It is during these phases that the formation and development of inclusive design and solutions are born and matured. It doesn't matter what phase of the design process is being worked, design lens is applied during each stage to

TABLE 4.1
Example Design Lens

1. Physiological
   - Weight
   - Height
2. Race
   - African American
   - Caucasians
   - Latinos
   - Asians
   - Jews
   - Native Americans
   - Pacific Islanders
   - Africans
   - Other
3. Sex
   - Male
   - Female
   - Bisexual
   - Homosexual
4. Disability (differently abled)
   - Speech
   - Physical
   - Visually impaired
   - Hearing impaired
5. Cultural attributes
   - Norms
   - Language
   - Artifacts
   - Symbols
   - Values
6. Religious beliefs
7. Inexperienced user
8. Financial capability
9. Knowledge gap and experience
10. Skin pigmentation
11. Aging population
12. User accessibility to product, system, technology
13. Traditions (regional, local)
14. Spoken language

produce optimal impact on a project or process. As the project team changes members, which will occur over time through the design process, new design lens will be introduced. Figure 4.2 show examples of when design lens could be applied; as shown, design lens can be applied and incorporated throughout the design and new product development process.

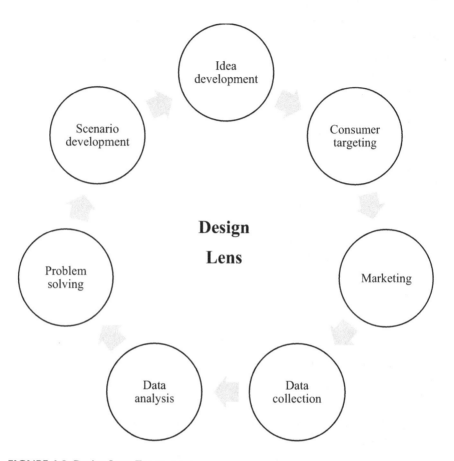

FIGURE 4.2 Design Lens Engagement.

When incorporating design lens into the design process, the designer or engineer should consider the following:

1. Evaluate the product or technology for a futuristic view
2. Product, process, technology, or services that are usable and safe
3. Enhance the quality of life for a diverse population
4. Provide an opportunity to visually see the technology in operation and how users will be impacted
5. Provide an inclusive consideration for consumers from all walks of life
6. Facilitate the exploration of alternate possibilities

4.2.1 Tips in Selecting a Design Lens

The primary application of design lens is to enact an inclusive framing for all design decisions and new product development. Therefore, it is critical that the new product or process development answer specific questions and address various aspects of the design process. Some questions that can assist in selecting the proper design lens are

- Who are the targeted or expected users/consumers?
- How will the product be used?
- Are there any populations that may be sensitive to the product or process?
- Is there a potential that a segment of the population may be in opposition?
- What are potential obstacles that can challenge or hinder process or product success?

4.3   VISION CONCEPTING

Corporate leaders are always seeking ways to provide the right product and services for consumers at the right time to address existing market needs and to sustain their businesses. To achieve this, they attempt to conduct forecasting scenarios to antic-ipate future needs and trends to proactively address those needs and get "ahead of the market." Oftentimes these forecasts are fed into a strategy that does not yield the expected outcome or quickly become outdated before implementation. Vision con-cepting enhances speculative thinking and attempts to sketch the future far beyond the traditional development time frame, up to 15 to 20 years into the future. **Vision concepting is defined as the identification and incorporation of societal factors into the design process.**

The creation of a vision concept assists in providing a view of what lays ahead, which can include factors such as the current competition, potential product per-formance, or even existing operational practices. Vision concepting enables current practices to be challenged and adjusted to identify a more probable and inclusive path for future product development and design. Although one of the primary rea-sons for creating vision-based concepts is to ensure the product design incorporates inclusivity into the development process, other benefits are realized. For example, the creation of product concepts requires collaboration among cross-functional team members, which brings different disciplines into diverse discussions on differences and solving problems. This interaction among the diverse team members will more likely improve communication and cohesion. In addition, working together to solve problems and develop concepts increases personal commitment to the company's objectives while engaging in creative approaches that facilitate development of new products, technologies, or systems.

Concepts can, and have been used, to turn abstract ideas and objectives into something more tangible. The idea of concepts usage formed the basis for a cross-disciplinary research conducted by Turkka and Roope, 2006. This research resulted in a method and process description for linking the generation of scenario, technology road mapping, and concept design that is used in generating vision con-cepts. This method, termed "vision concepting," consists of two primary phases in the methods that are future description and conceptualization. The two methods are divided into four categories identified as identification of change factors, scenario generation, identification of concept drivers, and generation of concepts. A summary of the Turkka and Roope, 2006 vision concepting process is shown in Figure 4.3.

FUTURE DESCRIPTION

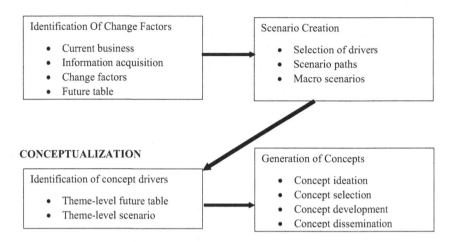

FIGURE 4.3 Vision Concepting.

4.4 A&D VISION CONCEPTING PROCESS

The vison concepting process devised by Turkka and Roope formed the basis of the vision concepting process devised by Alston and Millikin DeKerchove, referred to as the A&D Vision Concepting Process. The A&D Vision Concepting Process is presented in diagram Figure 4.4. As shown, design lens are applied throughout the A&D Vision Concepting Process.

The A&D Vision Concepting Process is based of four critical components that serve as a road map to help gather, analyze, and strategically deploy information that can be used in forecasting a future outcome and its impact on various products or products beginning at the discovery phase. These four components are discussed in further detail below.

It should be noted that the identification of change factors is performed through an intensive data gathering effort. This effort is necessary to identify and define change factors that will have an impact on an organization, product, or technology, a data gathering campaign. **Gathering data is a strategic process because information is sought that can impact future development and implementation of a design**. The information gathering process should take into consideration data and information which can have political, environmental, legal, social, technological, and economic impacts. The characteristics of the data that should be collected is dependent upon the design or technology seeking market introduction or modification. The data collection needs for the VCP are not as extensive when used with the 4-D Algorithm (Chapter 5) for new product design because the model has incorporated a data collection step within Phase 1 (discover) of the process. Vision concepting adds specificity in collecting data using design lenses that will facilitate inclusivity and a futuristic view of product or process evolution.

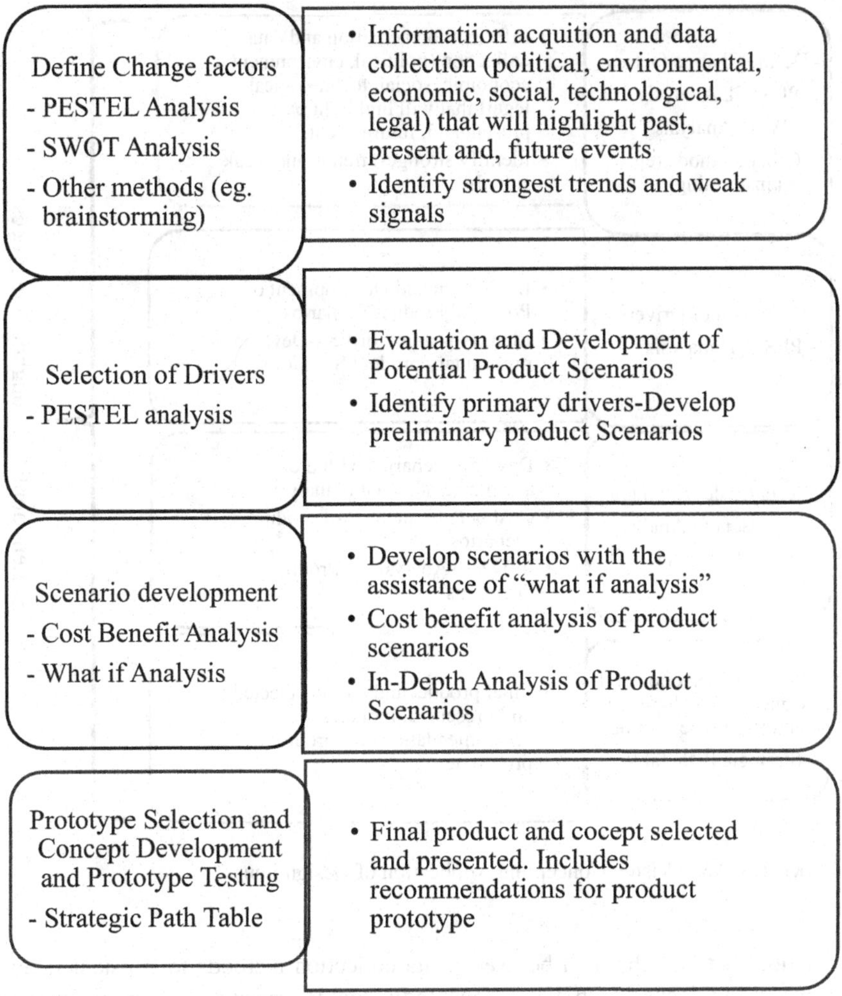

Define Change factors - PESTEL Analysis - SWOT Analysis - Other methods (eg. brainstorming)	• Informatiion acquition and data collection (political, environmental, economic, social, technological, legal) that will highlight past, present and, future events • Identify strongest trends and weak signals
Selection of Drivers - PESTEL analysis	• Evaluation and Development of Potential Product Scenarios • Identify primary drivers-Develop preliminary product Scenarios
Scenario development - Cost Benefit Analysis - What if Analysis	• Develop scenarios with the assistance of "what if analysis" • Cost benefit analysis of product scenarios • In-Depth Analysis of Product Scenarios
Prototype Selection and Concept Development and Prototype Testing - Strategic Path Table	• Final product and cocept selected and presented. Includes recommendations for product prototype

FIGURE 4.4 A&D Vision Concepting Process Components.

The information gathered should be organized and analyzed to determine political, economic, social, technical, legal, and environmental impacts while paying closing attention to critical change indicators. These change indicators can be helpful in modifying future plans and decision basis. Design lenses are applied throughout the process to ensure a focus on inclusivity is kept at the forefront throughout the vision concepting process (Figure 4.5).

4.4.1 Data Collection and Evaluation

Data collection and analysis are important elements that must be considered when implementing the vision concepting process. There are many collection and

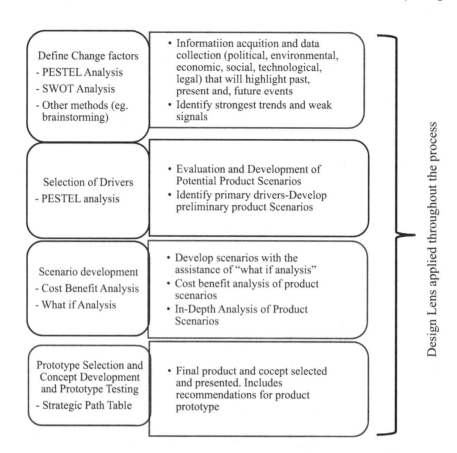

FIGURE 4.5 A&D Vision Concepting Application of Design Lens

evaluation methods that can be used. Data collection methods to consider include benchmarking, focus group discussions, individual interviews, and survey questionnaires. However, considerations must be given to the analytical methods that will be used before deciding on the collection methods.

Data evaluation can take on various forms, such as brainstorming, the political, economic, social, technological, environmental, and legal (PESTEL) analysis, "What-If" analysis, and the strengths, weaknesses, opportunities, and threats (SWOT) analysis. **The degree of analysis will vary depending upon the degree of acceptable risk.** There are other analysis methods that may be used in the A&D Vision Concepting Process; however, the authors chose the methods above because of their level of commonality.

4.4.1.1 The Political, Economic, Social, Technological, Environmental, and Legal Analysis Method

The political, economic, social, technological, environmental, and legal (PESTEL) analysis is a tool used to analyze and monitor macro-environmental (external

TABLE 4.2

Example PESTEL Factors

Factor	Description
Political	Factors include political stability or instability; political consequences; change in government constitutions; political and national policies; government actions; government stability; government support to organizations; exchange rate policies, and foreign trade policies.
Economic	Factors can include current and forecast economic growth, fluctuations in inflation and interest rates, job growth and unemployment, labor costs, disposable income of consumers and businesses, and projected changes in the economic environment.
Social	Factors may include demographics (age, gender, race, family size), consumer attitudes and behaviors, private opinions, purchase patterns, population growth rate, socio-cultural changes, ethnic and religious trends, and quality of living.
Technological	Factors that can influence marketing and new ways of producing goods and services, distributing goods and services, and communicating with target markets sectors.
Environmental	Factors such as those that are associated with scarcity of raw materials, pollution prevention targets, conducting business as an ethical and sustainable company, and carbon footprint targets.
Legal	Factors may include personal safety and health, equal opportunities, advertisement standards, consumer rights and laws, and product safety and labeling.

marketing environment) factors that have an impact on an organization, industry, technology, or product. It examines specific factors in the external environment. The model helps an organization to understand societal factors which can influence the market in which they compete and their position within that market. Example factors which influence the PESTEL analysis are listed and defined in Table 4.2.

The PESTEL analysis framework has been used effectively in macro-environmental analyses and in strategic planning for future uncertainties. **The PESTEL analysis can be used to group factors in the macro environment** (uncontrollable external forces that have an impact on the business, product, technologies, or services) **to allow the identification of opportunities and risks that have an impact on future outcomes**. Analyzing macro-environmental factors is particularly valuable when used to understand how external activities and events influence, drives, and difficulties impact on area of focus. The applicable data obtained from the PESTEL analysis can be further evaluated in a strategic planning table, as further discussed in 4.5 by listing the variables in rows and the potential state in columns.

4.4.1.2 Strengths, Weaknesses, Opportunities, and Threats Analysis

A strengths, weaknesses, opportunities, and threats (SWOT) analysis is a good tool to use for evaluation of a design and is often used to analyze the internal and external environments of organizations to aid in making the best decision on current and future product development paths. The four components of a SWOT analysis are designed to identify internal and external scenarios that can impact a design, organization, or product. The four components defined include the following:

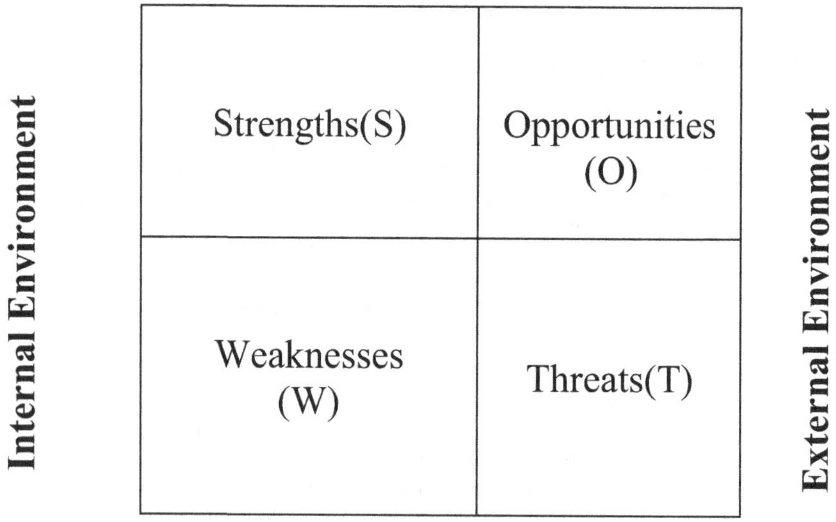

	Strengths(S)	Opportunities (O)	
Internal Environment	Weaknesses (W)	Threats(T)	**External Environment**

FIGURE 4.6 SWOT Diagram.

- Strengths—Internal elements of an organization that can facilitate goals and achievement.
- Weaknesses—Internal elements that interfere with organization's success.
- Opportunities—External aspects that help an organization reach its goals. These elements include positive environmental aspects and opportunities to address gaps and institute new activities.
- Threats—Aspects of an organization's external environment that are barriers or potential barriers to reach its goals.

The SWOT analysis has been used in many different disciplines, including education, industry, and agriculture. The SWOT model has been used in concert with other techniques, such as the PESTEL analysis, *analytic hierarchy process* (AHP), *and Five Forces Model*. These combinations have produced more accurate and powerful results and support strategic designs. The SWOT m is designed for, and effective in, assessing aspects of business in terms of strengths, weaknesses, opportunities, and threats. The SWOT analysis can be used effectively to produce alternative options (scenarios) for a business and design.

SWOT is a convenient tool at the evaluation stage to gain an initial idea of possible future consequences. The analysis is a simple yet effective method that can provide a realistic view of the strengths and weaknesses of a business. It provides the designer or engineer assistance in gaining an overview of the differences between actual and future plans and an analytical view of the current competition situation.

4.4.1.3 What-If Analysis

The "What analysis" is a simulation method with the goal of evaluating the potential behavior of a complex system under some given hypotheses called scenarios. Formulating scenarios enables the building of a hypothetical world or design that

TABLE 4.3

Example What- If Analysis for Hazardous Chemicals Storage

Item No.	What-If?	Consequences	Recommendations
1	Chemicals stored in storage room have not been classified.	Potential of storing incompatibles together and storage conditions discovered by regulators leading to potential fines	Classify all chemicals and store compatibles together
2	Containers are not completely sealed and affixed to contain content.	Mixing of chemicals can cause reactions creating hazardous fumes	Ensure that all containers are sealed properly
3	Limited walk space in storage rooms and containers	Personnel bumping into containers during routine inspections	Increase walk area to avoid encountering containers
4	Old containers are rusting and touching other containers.	Chemical reactions	Increase storage space for each drum and overpack rusting containers

can be analyzed and potentially provide a view into future activities and outcome. In a What-If hazard analysis, brainstorming techniques are used in the form of What-If questions to identify possible deviations and weaknesses or issues in design. Once identified, these concerns and their potential consequences are evaluated, risk determined, and mitigation attempted. Table 4.3 is representative of a What-If analysis for the storage of hazardous chemicals.

4.5 STRATEGIC PLANNING TABLE

A strategic planning table can be used to organize data to facilitate decisions, focus attention on scenarios that can assist in developing and implementing products that are usable for a diverse consumer base, and combine change factors into rational future paths. An example of a strategic planning table for a product to market launch is shown in Table 4.4. The table has been prepared for a Design Project XX, seeking to modify an existing technology being used to collect critical data to improve fitness. The current equipment is bulky and is only used in a medical facility or a physician' office. The modified unit will be small and compact and is anticipated to be worn on the wrist of individuals. The results obtained from data research have been cataloged into the strategic planning table in Table 4.4.

4.6 SCENARIO CREATION AND DRIVERS

Scenario creation may be viewed as the part of the vision concepting process that can potentially provide a look into the future and allow for changing directions early when changes arise and respond to market and customer uncertainties. **Steps in scenario development may include brainstorming of major driving forces or change factors in the development process, selecting drivers that impact change, rank forces by impact, rank forces by uncertainty, develop scenario logics, construct the research agenda, define the plots and titles, write the scenario stories, and develop**

TABLE 4.4

Strategic Planning—PESTEL Analysis Example

Design Factor	Political	Economical	Societal	Technological	Environmental	Legal
Cost	Trade agreements and tax tariffs may impact sales in other countries. Evaluated significance on marketing	May be driven by inflation; economic forecasts should be evaluated and factored into cost point of product.	The importance or value of fitness will vary among populations and societies.	Technology may not be cost-effective and readily available in some countries.	Eco-friendly products for material supply and packaging is desired.	Labor agreements, associated with product build and sell, should be evaluated and costs factored into overall product cost.
Product Distribution	Availability of product components is impacted by recent pandemic.	Cost and availability of distribution channels in countries external to the US	None identified	Degree of industrialization and technology may vary depending upon where products are built.	None identified	Impacts from labor agreements associated with product distribution
Raw Material	Country of production politically unstable and production material may not be available when needed	Produced only in third world country. Identify other production locale or alternative materials.	None identified	None identified	Materials used to build product are not associated with environmentally harming technology.	Some countries have specific legal agreements for material originating from their country.

TABLE 4.5

Theme level Strategic Planning Table for Project XX

Themes	Considerations for Team Solution Discussion
Supplies	Discussion and determination of all potential viable solutions that can be used to resolve current and potential supply impacts. Examples include contracting with various producers, developing the capability to produce the chip within the company, contracting and funding others to produce chip within the US.
People Resources	Evaluate and discuss ways to ensure that labor resources are available. For example, consider partnering with a local university to train potential workers in the discipline and competency needed, develop internal capability, retool job scope function to include the competency when and if needed.
Equipment and Parts	Explore various options for equipment purchase and maintenance. For example, procure a comparable part from different vendors, partner with local vendors.
Technology accuracy and acceptance	Explore factors that can impede operation and acceptance of the product as adding value to society and consumers. For example, can the product be used on different skin types, modification to current design to incorporate a wider usage, various body placement.

the scenario communication strategy. Creating scenarios that can impact the future of the design, technology or product is a means to predict potential occurrence and consider what solutions may be implemented to address anticipated issues. It also provides an opportunity to potentially make modification in the design when practicable.

When developing scenarios, consideration must also be given to the impact of environmental, political, social, legal, economic, and technological impacts into the future. To assist in developing the appropriate scenarios it is recommended that the What-If analysis process be used. Scenario drivers may be used in the development of a strategic planning table, and the scenarios previously developed can be further catalogued into a theme level table with associated theme scenarios. Table 4.5 presents an example of a theme level table. The theme level scenarios are scenarios that have been made specific for a particular concept or design. Formulated themes may be used by the design team to

- Validate and verify the proposed design, technology, or product
- Determine changes that can be made to enhance the product
- Predict potential challenges and develop actions that can be taken to eliminate or reduce impacts
- Optimize design or product and introduction to market

4.7 PROTOTYPE SELECTION AND CONCEPT DEVELOPMENT

The previously generated scenarios and themes generated may be instrumental in stimulating conversation with the design team to fully flush out and generate additional ideas for the design, product, technology, or services being explored, developed, or modified. **In generating concepts, the focus of concept ideation is to stimulate and develop ideas using previously generated theme scenarios for the product or**

	Strengths	Opportunities	
Internal Environment	Strong technology	Invest in chip manufacturing	**External Environment**
	Weaknesses	**Threats**	
	People resources	Chips attainment Resource needs COVID 19 continuance	

FIGURE 4.7 SWOT Diagram for Project XX.

process and the applicable technology, service, and product road map (strategy).
It is during this process that the concepts generated are flushed out, ranked, grouped,
or combined, yielding a final list of concepts that can be linked to the generated scenarios. The draft list of concepts having the most potential are further developed
throughout the design process. At this stage the final design of the specific technology,
product, or services along with the market specifications are finalized. At the end of
this process, it may be helpful for the team to perform a SWOT analysis to ensure that
pertinent aspects have been identified and addressed. A simplistic SWOT analysis for
Project XX may resemble the analysis shown in Figure 4.7.

4.8 SCENARIO PLANNING AND GENERATION

A scenario is defined by Oteros-Rozas et al, 2015 as a coherent, internally consistent,
and plausible description of a potential future trajectory of a system. Within this
book the authors have modified that definition to read: "A scenario is a coherent,
consistent, and highly feasible depiction of a potential future trajectory of a project,
technology, system, or process." The scenario planning processes are often oriented
toward influencing decisions that can potentially have a wide range of implications
for a diverse set of stakeholders.

Scenario planning can address the following:

1. Synthesize information concerning what is important for an organization,
   technology, project, or process, a necessary foundation for understanding
   future uncertainties.

2. The development of a consistent and likely set of descriptions of possible futures or scenarios, using a structured methodology.
3. The evaluation of the implications of these scenarios for the organization, technology, product, project, or today.

The process of scenario planning involves four primary stages. These stages are detailed below.

- Stage 1: Setting the Scene. Defining the purpose of the planning exercise, developing an understanding of the current instate situation, set a time horizon, select the appropriate participants, and define the need for the scenario planning process which normally takes place as a preparatory activity
- Stage 2: Identifying key driving forces. This can be accomplished through interviews of key stakeholders or within a workshop setting. Although there will be some variation in practice, the driving forces that shape the future should concern the general environment following PEST or other equivalent analysis. This stage may be accomplished within a workshop setting with brainstorming session. Considerations should be given for preparing a series of key questions to be used to help inform and guide interviews.
- Stage 3: Ranking forces by the level of uncertainty and impact. This can be accomplished by using a two-axis diagram to evaluate the relative importance and level of uncertainty for each factor in a qualitative, discussion-based approach. This diagram is used to assemble the driving forces identified in the previous stage to select the most important uncertainties. It is suggested that the potential maximum and minimum values of each of the selected uncertainties should be considered.
- Stage 4: Selecting central themes and developing scenarios, using various techniques depending on the contextual setting of the exercise. The goal is to develop plausible scenarios from the information collected, realizing that there is a lot of flexibility in how this stage is accomplished.

Note that there is significant variation between different studies concerning how many scenarios should be identified. The recommended number of scenarios to be developed should be between two to eight. There are inductive and deductive methods that can be used to identify the scenarios' themes. The inductive approach is based on building the scenarios around uncertainties, and the deductive approach is based on pairing two uncertainties from those selected in the previous stage to create four alternative scenarios. Once the themes and number of scenarios have been developed, the scenarios should be developed and written into a narrative form. Why use scenarios?

1. To stimulate or provoke strategic conversations during the design process
2. To stimulate new ideas and visionary thinking during the design process
3. As a motivator and method for getting "unstuck" during the design process

4.9 FUTURE THINKING CONCEPT AND TOOLS

Future thinking is comprised of several facets of cognition that include problem-solving, judgement, decision-making, self-reflection, and the ability to anticipate future events based on past experiences. Thinking is greatly influenced by one's ability to remember the past, the representation of knowledge, and one's perception of the world. Thinking and memory are strongly interrelated, and problem-solving and thinking are inseparably linked together.

An important function that combines memory and thinking is mental time travelling, commonly referred to as thinking about the future. The importance of the ability to anticipate the future based upon the past experiences has been interestingly expressed by Schacter et al. (2007, p. 660) in the paragraph entitled "The prospective brain," in which he writes "preparing for the future is a vital task in any domain of cognition or behavior that is important for Thinking and Memory that is important for survival."

Futures thinking is an approach of considering, and hopefully, shaping the way that systems, technology, and relationships will develop over time. Futures thinking requires cautious and reflective analysis of current conditions, pressures on the system, risk, resources, and current trends. This analysis will make visible those futures that are possible and beneficial as well as what futures seem impossible or undesirable. It is important that futures thinking be grounded in the present and be well informed by empirical data about the current state of things.

There are many activities that can, and are used to, facilitate and foster reflections into future thinking that include tracking trends, scenario statements, future wheel, and black swans. These activities were used by Hinchliffe during a Future Thinking Workshop and are briefly summarized in Table 4.6.

TABLE 4.6
Future Thinking Activities

Activity	Activity, Steps, and Parameters
Trend Analysis	(1) Identify and distribute an analysis of data to participants/team members
	(2) Guide the discussion using a series of scaffolding questions, emphasizing comprehension, and expanding into analysis, critique, and implications
	(3) Summarize the discussion and actions generated
Scenario Statements	Use short scenario statements which summarize current trends or a descriptive statement of the current state of an object. Reflective questions may be used to prompt and guide reflections and questions.
	(1) Develop scenario statements
	(2) Divide participants into small groups for discussion
	(3) Ask each group to prepare a summary of discussion and present. Ask listeners to take notes of common themes and unique ideas presented
	(4) Facilitate a reflective discussion of themes highlighting implications, challenges, etc.
	(5) Summarize the discussion and all actions generated

Activity	Activity, Steps, and Parameters
Future Wheel	Develop a visual method to explore implications of a change through brainstorming consequences of the change and other potential consequences. The resultant product is a map of consequences referred to as a futures wheel.
	(1) Provide an example futures wheel to be examined and reflected by participants
	(2) Divide participants in small groups and assign each group to place a change at the center of the wheel
	(3) Order consequences from first order to third order
	(4) Facilitate reflective discussions
	(5) Summarize discussion and potential actions
Black Swans	A change that would be viewed as surprising that would have significant impact on the future. These changes are difficult to predict.
	(1) Provide a black swan statement for considerations by participant
	(2) Divide participants in small groups
	(3) Request each group prepare a summary of their discussion and present summary to the entire group requesting listeners to take notes on common themes and unique ideas
	(4) Facilitate reflective discussions of themes and potential implications
	(5) Summarize discussions and potential actions

4.10 APPLIED LEARNING

Read and respond to the questions below. Your response should be in the form of an essay. Conceptual questions have been presented to assist in providing a comprehensive response on each topic.

1. Explain the concept of design imagination. How can this concept be used to discover and develop new products and technologies to meet the needs of a progressive technological society and improve the life of the human population?

2. What are design lens? How are they used in product and technology development? List at least five design lenses and how they can be used in technology and product development? How can a designer select the design lens that will help expand the customer base for a product or technology? What are the potential dangers of not using design lens or selecting the wrong lenses? What are the potential benefits of using design lenses?

3. Define Vision Concepting. Define the components of the A&D Vision Concepting model and practical usage. How can and should this concept be used in providing and introducing new products and technologies? What data collection strategies should be used to support the Vision concepting usage? What are some of the data analysis methods that can be used to assist in providing value to decision-making for management?

BIBLIOGRAPHY

Benzaghta, M. A., Elwalda, A., Mousa, M. M., Erkan, I., & Rahman, M. SWOT analysis applications: An integrative literature review. *Journal of Global Business Insights*, 6(1), 2021, 55–73. www.doi.org/10.5038/2640–6489.6.1.1148

Chermack, T. J. *Scenario Planning in Organizations: How to Create, Use, and Assess Scenarios*. Berrett-Koehler, 2011

Deis, R. Stefano, What if Analysis, University of Bologna, Italy, 0000836408 1.5 (unibo.it), 2–5–2008

Folkmann, M. N. *The Aesthetics of Imagination in Design*, MIT Oress, 2013 (Proquest Ebook Central)

Folkmann, M. N. Unknown positions of imagination in design. *Design Issues*, 30(4), 2014, 6–19. https://doi.org/10.1162/desi a 00293

Glatzeder, B., Goel, V., & Mülle, A. *Towards a Theory of Thinking: Building Blocks for a Conceptual Framework*, Springer, 2012. www.springer.com/us/book/9783642031281)

Hinchliffe, L. J. The futures of scholarly communications: techniques and tools for futures thinking. *The Serials Librarian*, 78(1–4), 2020, 28–33. https://doi.org/10.1080/03615 26X.2020.1739473 https://libguides.libraries.wsu.edu/c.php?g=294263&p=4358409

Hussain, M., Tapinos E., & Knight, L. Scenario-driven roadmapping for technology foresight. *Technological Forecasting and Social Change*, 124, 2017, 160–177. https://doi.org/10.1016/j.techfore.2017.05.005

Keinonen, T., & Takala, R. *Product Concept Design: A Review of the Conceptual Design of Products in Industry*, Springer, 2006.

O'Brie, F. A. Scenario planning—lessons for practice from teaching and learning. *European Journal of Occupational Research*, 152(3), 2004, 709–722. https://doi.org/10.1016/s0377-221(03)00068-7

Ogilvy, J. A. *Facing the Fold: Essays on Scenario Planning*, Triarchy Press, 2011.

Oteros-Rozas, E., Martin-Lopez, B., Daw, T. M., Bohensky, E. L., Butler, J. R. A., Hill, R., Hanspach, J. Participatory scenario planning in place-based social-ecological research: insights and experiences from 23 case studies. *Ecology and Society*, 20(4), 2015. https://doi.org/10.5751/es-07985-200432

Ralph, N., Birks, M., Chapman, Y., & Francis, K. *Future-Proofing Nursing Education: An Australian Perspective*, SAGE Open, 2014, pp. 1–11. https://doi.org/10.1177/2158244014556633 sgo.sagepub.co

Rashain, P. *The PESTLE Analysis*, Nerdynaut, 2017

Schacter DL, Addis DR, Buckner RL. Remembering the past to imagine the future: the prospective brain. Nat Rev Neurosci. 2007 Sep;8(9):657–61. doi: 10.1038/nrn2213. PMID: 17700624.

Sivapirakasham, S. P., Thiyagarajan, S., Bineesh, P., Mathew, J., & Surianarayanan, M. Hazard identification in electrical discharge machining (EDM) process using what-if analysis. *Applied Mechanics and Materials*, 592–594, 2014, 2508–2512. http://dx.doi.org/10.4028/www.scientific.net/AMM.592-594.2508

5 Vision-Enabled Design Thinking

Incorporation of Design Lens and Vision Concepting into the Design Thinking Process

5.1 UK DESIGN COUNCIL: DOUBLE DIAMOND DESIGN PROCESS

The UK Design Council was established in 1944 with the objective "to promote by all practicable means the improvement of design in the products of British industry." Since 1944 the organization and approach to design has evolved, transitioning their focus to technology and then engineering. As an awareness of design among businesses increased in the late 1990s, the Design Council recognized a change in focus on how design was needed. This led to a shift in the Design Council's purpose to include enabling, as well as inspire the use of design.

In 2004 the UK Design Council published the design methodology, the Double Diamond, as a comprehensive and visual description of the design process. Figure 5.1 depicts the original Double Diamond design process. As illustrated, **the Double Diamond design process can be used as an approach to design but also to problem-solving, which is indicative of the thought process used when designing products or even services.**

The two diamonds represent a process of exploring an issue more widely or deeply, using divergent and convergent thinking, and then taking focused actions using convergent thinking. The Double Diamond design process is based on progressing through the four phases of design which are further discussed below. Although not shown in Figure 5.1, one of the key facets of the Double Diamond design process is the recognition it is not a linear process; rather it is an iterative process. As the designer or engineer progresses through the design process, they may have a need to revisit an earlier stage of the process, such as back to the discovery stage.

> **Phase 1: Discover.** Understand, rather than assume, what the problem is. Speaking to and spending time with people who are affected by the issues. The discovery phase is a designer or engineer's information gathering phase. Essentially, this phase is used to understand feedback and needs of potential customers and

DOI: 10.1201/9780367854720-6

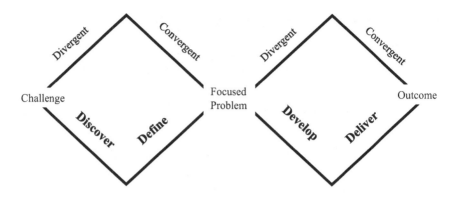

FIGURE 5.1 Double Diamond Design Process.

uses divergent thinking to gather as much information as possible by different means or mechanisms, from as many sources as possible.

Phase 2: Define. Insight or information gathered during the discovery phase is then evaluated and analyzed to define the challenge or problem in different ways. The define phase is implemented by the designers or engineers by analyzing data to identify several problems or challenges to be further evaluated. Convergent thinking is used to analyze and "envelope" the problem or scenarios to move forward for further development and analysis. Criteria are often defined at this point in the design process, by which to perform the more in-depth analysis in Phase 3.

Phase 3: Develop. The focused problem or scenarios identified as warranting further analysis are evaluated against specified criteria to finalize conception of the problem or product/preferred scenario. Divergent thinking is applied to introduce different approaches and ideas to analyze and refine a selected product or problem resolution. How the product or design would be applied are discussed, along with understanding who is the customer. Prototype testing can be employed in this phase, as well as pilot studies with different target audiences to better understand what humanistic needs are being met through use and application of this product or problem resolution.

Phase 4: Deliver. The deliver phase may include additional pilot-scale testing and takes into consideration the development of a marketing campaign that uses design thinking to target audiences. Convergent thinking is used to better understand how to improve deployment of the new product or problem resolution. Cost can be a significant driver in delivery or deployment of the new product because of limited dollars available to execute a new product.

The key to effectively applying the Double Diamond design process is the robust use of divergent and convergent thinking.

5.1.1 Divergent and Convergent Thinking

When performing human thinking, there are three modes: lateral, divergent, and convergent thought. The terms divergent and convergent thinking are used to communicate the approach people use to think. Experience has shown that teams often use these two methods together to achieve the best results.

Divergent thinking is the thought process or approach used to generate creative ideas by exploring different solutions or resolutions. Divergent thinking is often described as "non-linear," such that ideas are spontaneous, non-structured, and generated in an emergent cognitive fashion. A great example of applying the use of divergent thinking is when you take a pile of blocks and stack them in a manner by which you can create different designs. Once you have created as many different stack designs as possible, you then apply convergent thinking to organize and structure the stacks until a final stack design is achieved.

Convergent thinking is the thought process or approach used to analyze and determine what is the appropriate or correct answer to a need or problem. Cognitive thinking does not require creativity; rather, cognitive thinking is linear and reproducible through a specific thought process. One of the best examples of convergent thinking is someone taking a driver's license test; there is only one correct answer to each test question. When taking the driver's license test, a person uses critical thinking that is linear and logical when analyzing the driving solution to a problem and to identify the correct answer. Convergent thinking is applied in the Double Diamond design process when analyzing ideas against criteria or defined needs that an idea is addressing.

There is a third thinking mode which is lateral thinking. Lateral thinking is often described as "outside the box" thinking. The use of lateral thinking is creatively driven and often is used when a radical change is desired. One can also use lateral thinking for envisioning the future of a product or service. Lateral thinking involves indirect and creative thinking that in many applications is not intuitively obvious. Lateral thinking is not incorporated into the Double Diamond design process; however, lateral thinking may be used in conjunction with divergent thinking to gather a larger set of ideas to analyze.

5.1.2 UK Design Council *Framework for Innovation*

Over time the UK Design Council recognized that the Double Diamond design process was not developed as a linear process; rather, progression through each phase of the model (i.e., define—Phase 2 and develop—Phase 3) could, at times, lead you back to an earlier model phase (i.e., Discover—Phase 1); organizations discovered the model assisted in learning about underlying problems that would drive them back to the discover phase. In 2019, the UK Design Council established the *Framework for Innovation*, which incorporated key principles and methods that designers and non-designers use to achieve significant and long-lasting positive change. Figure 5.2 depicts the UK Design Council *Framework for Innovation*. Although not shown on Figure 5.2, similar to the Double Diamond design process, it is not a linear process; rather it is an iterative process. As the designer or engineer progresses through the

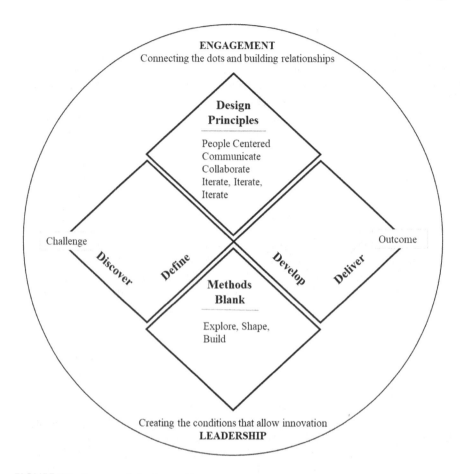

FIGURE 5.2 *Framework for Innovation.*

design process, they may have a need to revisit an earlier stage of the process, such as back to the discover stage.

The UK Design Council *Framework for Innovation* incorporated design principles for designers or engineers which included the following:

- **People first**. Designers and engineers should understand the people they are serving, their needs, strengths, and aspirations. What drives the clients you are serving and include those characteristics into the product or problem.
- **Visually and inclusively communicate**. The designs should solicit feedback in different ways whether it be visually or incorporate inclusivity into the design or solution to better fit the needs of the client.
- **Collaborate and co-create**. Collectively we are stronger than individually as we design. Working together can inspire and drive others to create in a manner they normally would not have created.

- **Iterate, iterate, and iterate**. To iterate means to build knowledge through experimentation, refinement, and repeatability. The iteration process is used to identify errors and build upon resolution for improvement and recognizes the need to sometimes return to Phase 2 of the design process, while implementing Phase 4.

Two of the more important concepts that were introduced by the UK Design Council into the Double Diamond design process was leadership and engagement in the overall design process. Leadership, whether it is organizational or individually driven, establishes the vision, expectations, and acceptable behavior associated with the application of the design process within the organization or company. Different leadership styles are further discussed in Chapter 9 of this book, but leadership characteristics that the *Framework for Innovation* identify include

- Encouragement to innovate
- Enhance skills and capabilities
- Acceptability of experimentation and learning
- Adaptation is the norm and understanding results and learning from them

Engagement of stakeholders, and other parties, who may have an interest in the outcome of the design process was also recognized as a significant contributor to influencing the outcome of the product.

In today's business environment, design problems faced by organizations worldwide have become multifaceted and are a part of increasingly complex business models across the world. The expansion of global transactions, growth of international partnerships and the decentralized base of resources have led to challenges that require a global outlook and hence, an adaptation of the design process.

5.2   VISION-ENABLED DESIGN THINKING

As discussed in Chapter 2, innovation is at the heart of humanistic design thinking. **Vision-Enabled Design Thinking (VEDT) is an approach to design thinking that incorporates design lens and vision concepting into the design process to empower the designer or engineer to visually construct a design solution.** VEDT is an approach to visualization through a design framework that focuses on thinking in the future and uses a human-centered approach to speculative design.

The concept of VEDT was introduced by Woodrow Winchester in 2019, *Inclusive and Consequential by Design: "Futurefying" New Product Development (NPD) Through Vision Concepting,* where it was recognized that system engineers engaged in new product development manage technological considerations but also sociocultural consequences of their practices and outcomes and solutions. As we progress through the 21st century and as the world continues to merge into one global technological market, the ability to understand the impact of societal factors and diversity on new product design and development becomes critical to expand market capability.

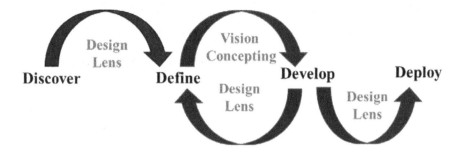

FIGURE 5.3 VEDT Applied to the *Framework for Innovation* Phases.

VEDT is an approach to visualization through a design framework that focuses on future thinking and drives consequential and anticipatory design thinking which leads to more inclusive designs. By implementing VEDT, through the application of design lens and vision concepting, diversity and inclusivity is promoted through incorporation of unique human-centered factors which can drive future product development.

The concept of VEDT was developed based on an adaptation of the Double Diamond design process and the incorporation of vision concepting. One could say that design lens and vision concepting are integral parts of VEDT. Figure 5.3 depicts VEDT, implemented through the A&D Vision Concepting process (VCP) as applied to the UK Design Council *Framework for Innovation* design phases. Note that implementation of each phase is in alignment with traditional phases of product development with the inclusion of considerations for designing products that will also focus on inclusion and end usage by consumers. Critical attention should be paid to Phases 2 and 3 because within these steps is where a predominance of humanistic design considerations is enabled and steps are taken to evaluate, consider, and ensure inclusivity.

The application of VEDT provides a vehicle for the designer or engineer to develop the design and communicate it in a language that is easily understood, and hopefully desired, by the customer, thereby promoting development of the visual concept in an efficient manner because the designer or engineer understands how a customer thinks based on the influence of societal factors. Through application of societal factors, designers and customers can better visualize and recognize how a design or product will address their current and future needs.

5.2.1 IMPLEMENTATION OF VEDT WITHIN THE FOUR DESIGN PHASES

Implementation of VEDT includes the application of vision concepting in primarily the second and third phase of the UK Design Council *Framework for Innovation*. As the designer or engineer progresses from the define to the develop phase, unique design lenses and vision concepting are applied and iteratively considered and discussed, which help mold and shape the analysis of information and starts the process of visualizing how a product could be customer focused for maximum benefit.

The application of design lens and vision concepting truly humanizes the design through the speculative design process.

As discussed in Chapter 4, the A&D VCP consists of four components and may use various analysis tools, such as brainstorming, PESTEL, and SWOT to incorporate society factors into the design process through the integration of design lenses. Examples of societal factors which can be considered when developing tools include

- Customs and traditions
- Environmental justice
- Income and wealth
- Race
- Religion
- Rights and freedoms

Societal factors can include cultural norms and affect our attitudes, opinions, interests, and acceptance of a product (and design). Through each unique design lens, the application of societal experiences and perceptions of the design team will drive a visual concept or solution.

The identification and selection of societal factors are influenced by the new product being imagined and developed or problem being solved. **Design lenses are based on individual beliefs, experiences, and perceptions of the individual designer or engineer and are applied through all four phases of the A&D Vision Concepting Process.** Figure 5.4 identifies the A&D Vision Concepting Process components, as defined in Chapter 4, cross walked to the four phases of the *Framework for Innovation*.

A description of each of the four phases and application of the A&D VCP is further described below.

Discover	Define	Develop	Deploy
•Developing discovery tools •NPD project strategy development and execution	•A&D Vision Concepting Process applied in the analysis of information •Defining Change Factors •Selecting drivers	•A&D Vision Concepting Process applied to organize and refine •Scenario Development •Prototype Selection and Concept Development	• Development of marketing tools and execution •Incorporate feedback for future product development

FIGURE 5.4 Applying Vision Concepting to the *Framework for Innovation* Phases.

Phase 1 — Discover. As the design manager embarks on the new product design process, the first phase is discovery of information that can be used to develop concepts and ideas for addressing a need (new product) or enhancing an existing need currently being met (next product generation).

When developing tools used to gather the information, consideration should be given to aspects or factors of society which could be relevant, so the designer and engineer can incorporate that information into the data gathering method. For example, if focus groups are used to understand how a current product is perceived and useful to customers, then questions can be incorporated into the focus group inquiries, which will drive an understanding of the meaning, purpose, and priority of a societal issue to the potential customer.

Information obtained from focus groups vary depending upon the location and history of societal issues upon that customer population. The information can be useful in further analysis and interpretation of vision concepts applied in the next design phase but also when executing a marketing strategy for new product deployment. It is at the discovery phase that project management of the new product design process begins. How much time and money will be used to investigate and analyze societal factors in the overall new product design and development?

Phase 2—Define. Once information has been gathered, then analysis and refinement of the data begins. Information can then be organized and vision concepting tools applied, and through the application of design lens, analyzed to define criteria and potential visions of product scenarios. Idea formulation and development are initial points where a product takes on some form or shape. This phase takes place as an initial thought that a designer or engineer may have after experiencing or becoming aware of a product that is perceived as being incomplete or not functioning to its maximum capacity or perhaps feedback from colleagues, friends, or family member of the failures of a product. Regardless of how the formulation of ideas began, this phase may be viewed as the most important because without ideas there will be no new product developed. The phase of idea formulation and development process is dynamic, iterative, and interconnected, having the following characteristics:

- Idea and concept not set in stone at the onset
- Frequent flushing out until stability is achieved
- Requires repetition to explore validity
- Evaluation of idea many times—second guessing the concept

The application of vision concepting, and design lens, can be used not only in refinement of the design, but also to stimulate engagement from the design team. Every person wants to feel as though they matter and can add value when creating a product, and through open communications and applying vision concepting tools each team member can openly express their opinions and experiences.

Phase 3—Develop. The characteristics of this phase are closely aligned and inter-connected with Phase 2. This means that in order to finalize an idea that will lead to a design that produces a new product, all elements of idea formulation and development will play a significant role throughout the phase at the same time or individually. **As the vision of the new product becomes realized, vision concepting is used to further evaluate and refine preferred product scenarios**. The identification of scenario drivers and themes can be useful in development of prototype and trial deployment. Often designers or engineers will return to tools used in Phase 2, and sometimes Phase 1, when refining and developing a prototype and test market strategy. The design team continues to apply their design lenses to the iterative design process. Application of vision concepting at this phase drives to design inclusivity when proposing final product design and production.

Phase 4—Deploy. Insights gained throughout the design phase process are applied and used to leverage an effective marketing strategy that is inclusive to a target customer population. Certain features of the new product that appeal to the target customer base are emphasized through marketing that is tailored to a set of societal factors. Designers or engineers may refer to the previous three design phases to gain insight into marketing campaigns. Final product selection may occur during Phase 3 or 4, but societal factors should have been incorporated into the design so that the product appeals to a target customer population. Additionally, scenarios not used and vision concepting information associated with scenarios that were not carried forward, are stored for future product development and marketing activities.

The authors have developed the 4-D Algorithm for New Product Design as a more linear approach to applying VEDT into new product design. **The linear approach to design takes into consideration financial constraints associated with product design in a competitive business market, along with project management skills into the speculative design process, since designers or engineers are often responsible for managing design projects.**

5.3   4-D ALGORITHM FOR NEW PRODUCT DESIGN

The 4-D Algorithm for New Product Design is essentially a linear approach to the UK Design Council's *Framework for Innovation* and incorporates VEDT through the application of vision concepting and design lens, along with basic project management skills which ensures success from design origination to full deployment. As depicted in Figure 5.5, **the 4-D Algorithm for New Product Design was developed based on the fundamental premise that designing a new product must be effectively managed and should be an integral part of the business portfolio.** As part of the business portfolio, new product development is managed as an organization with defined projects that are managed through the use of basic project management skills.

The steps of the 4-D Algorithm for New Product Design originate from the UK Design Council's Double Diamond design process, but have been modified and sequenced in a linear fashion, recognizing the significant iterative design process on concept development

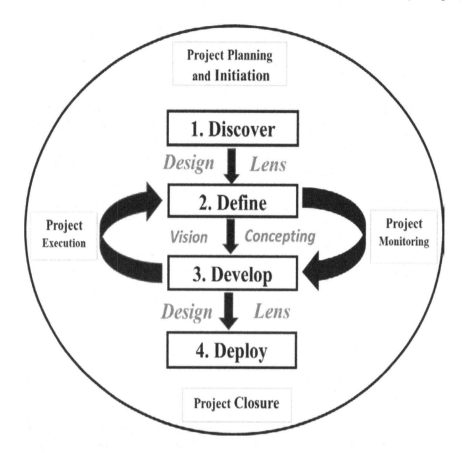

FIGURE 5.5 4-D Algorithm for New Product Design.

between Steps Two and Three. The 4-D Algorithm for New Product Design is different from the UK Design Council's Double Diamond design process in the following unique characteristics, which are aligned with current business practices:

- Based on sequentially performing the steps versus an iterative process
- Incorporates speculative design, design lens, and VEDT through vision concepting for greater inclusivity
- Incorporates fundamental project management principles into the process

Each of these characteristics are further described below.

5.3.1   4-D Algorithm and Sequential/Iterative Performance of the Steps

The UK Design Council's framework for innovation was based on four phases of the design process. The originators of the Double Diamond design process recognized that design is not a linear process; rather, it is an iterative process. Thus, the design process continues to mature over time. However, the realities of today's business environment are such that every department, including new product design and development, has a defined budget. As depicted in Figure 5.5, the 4-D algorithm for new product design recognizes the iterative design process that takes place between the four design steps; however, once a design is developed, the impact of being iterative is limited because most companies have a defined budget for product design and development. The need for a company to invest dollars into new product design is driven by the financial incentive and gains associated with product execution, deployment, and projected financial gains. In addition, design team members with special expertise (e.g., business marketer) will be needed for deploying the product versus design development, so your design team may also change depending on the design step. **Implementation of the design process through the 4-D algorithm for new product design drives performance through a number of sequential series, but also recognizes that design is an iterative process and must be effectively managed to continue to add value in today's competitive business market.**

5.3.2 INCORPORATION OF VEDT INTO THE SPECULATIVE DESIGN PROCESS

Implementation of the design process through the 4-D Algorithm for New Product Design recognizes the need to incorporate inclusivity into the speculative design process. As discussed in Chapter 3, speculation in design can be summarized as being a theory-based practice, where conjecture, abstract reasoning, and intellectual speculation is used to develop ideas into designs that advance technology and products while shaping the future for consumers. Through the application of VEDT, the 4-D Algorithm for New Product Design uses the A&D Vision Concepting Process to incorporate inclusivity into the design while applying abstract reasoning and speculation of future product need. Activating the imagination is an important aspect of successfully implementing the 4-D Algorithm for New Product Design. The entire process is built upon envisioning and understanding what consumers will desire in the future, based on historical and future artifacts and team members application of their design lens to address their basic and supplemental human needs. As written by writer and philosopher George Santayana, "Those who cannot remember the past are condemned to repeat it." Through application of the A&D Vision Concepting Process, historical and future considerations such as economic, political, and cultural aspects of society that are highly influential in the daily lives of people are identified and evaluated for application in implementation of the 4-D Algorithm for New Product Design.

5.3.3 INCORPORATION OF PROJECT MANAGEMENT PRINCIPLES INTO THE DESIGN PROCESS

All businesses, whether they are large or small, operate to a budget, and typically included in the budget is an organization dedicated to research and development, which may include new product design. The organization will have a defined operational budget that must be taken into consideration when envisioning existing product improvements or new product designs. The 4-D Algorithm for New Product Design was specifically designed to incorporate project management principles into the overall design process, but to also align with how projects in business are funded.

Project management is the process of planning, organizing, controlling, and directing resources, for a defined period of time, in order to complete a defined goal. All companies utilize project management fundamentals and company leaders, including management of new or improved product designs, must manage their organization and company within defined financial constraints. The concept of project management is a proven process that can be effectively used to manage the overall design process.

Fundamental components of project management include scope, cost, and schedule for completion of each step of the 4-D Algorithm for New Product Design. Typically, there are four stages when executing project management: project planning and initiation, project execution, project monitoring, and project closure. Figure 5.6 depicts the flow of activities used for project management.

When defining the scope of either a new or improved product design, each step and scope of the 4-D Algorithm for New Product Design must be taken into account and planned for, as funding will be sought for the total project upfront. As depicted in Figure 5.7, project management begins before the design

FIGURE 5.6 Project Management Activities.

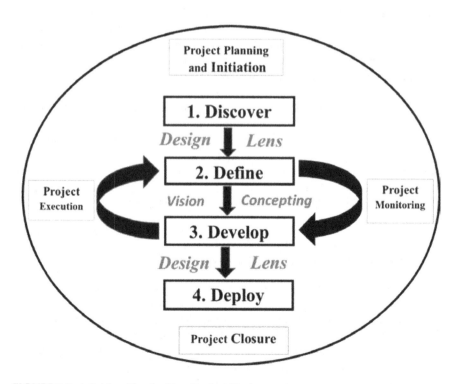

FIGURE 5.7 4-D Algorithm for New Product Design.

TABLE 5.1

Project Stages and Example Tasks

Project Stage	Example Task
Planning and Initiation	Develop the project scope
	Determine team resources, subject matter experts, and needs
	Develop budget, including any sub-elements
	Develop a communication and risk mitigation strategy
	Develop project/work plan
	Develop project schedule
	Obtain management approval for funding of the project/NPD
Execution	Ensure project team has required knowledge
	Direct and lead the project team
	Communicate key aspects of work execution to key stakeholders
Monitoring	Measure and monitor project performance against the project plan
	Manage project risks and challenges—which can include both technical and human
	Conduct routine status review meetings
	Document lessons learned along the way for future design projects
Closure	Obtain feedback for project usability
	Resolve any outstanding discovered issues
	Document design artifacts and electronically store for future design efforts

process is executed because companies need to be able to submit a documented project plan to obtain funding. Once approval and funding for the project has been obtained, then Step 1, discover, can begin. The project is executed as the design team progresses through each step of the 4-D Algorithm for New Product Design, as well as the budget and schedule monitored, to ensure the design is evolving and maturing to a level by which the design can be built and market tested. As the design process is an iterative process, the 4-D Algorithm for New Product Design recognizes and promotes design imagination and iteration while managing the design process to be competitive, and of greatest value to the customer. Table 5.1 identifies example tasks which may occur during each stage of project management.

The lead designer or engineer on the project may act as the project manager or there may be a dedicated manager, such as a section or division manager, who is responsible for multiple project scopes, cost, and schedules.

Project scheduling is an integral part of managing new product design as it provides the means to transform the project from a vision to a well-thought-out plan, as it provides the means to track progress, resource needs, and expenditures needed to make the design process a success. Because of the iterative nature of the design process, the project schedule should be regularly reviewed and realigned as needed. Not only is the design process iterative, management of the project scope and schedule is iterative as well, and by routinely managing the project schedule, the lead designer or engineer will ensure overall project success.

5.4   APPLICATION OF THE 4-D ALGORITHM FOR NEW PRODUCT DESIGN IN TODAY'S BUSINESS ENVIRONMENT

Today's business markets and clients are no longer thought of as only within one country or culture; rather, the world is viewed as one global market. The accessibility and use of the intranet for selling goods and services online has cemented the ability of companies to reach a greater number of consumers with their product. Many companies are re-evaluating the means by which their products are distributed, such as in physical retail stores versus online sales, to improve profits and market competitiveness. To remain competitive, companies are continuously re-evaluating not only products and services offered to their customer base, but companies are also looking for approaches on how to build and grow a diverse design team and/or organization. **The use of speculative thinking in the design process should take into consideration design artifacts, but also be reflective of historical successes and failures**. As designers or engineers progress through steps in the 4-D Algorithm for New Product Design and apply vision concepting and project management, the design process will mature and be cost-effective.

5.5   APPLIED LEARNING

1. What is the framework for innovation?
2. What is divergent and convergent thinking and how is it applied to the framework for innovation?
3. Explain VEDT and how it can be applied in the framework for innovation?
4. You have been selected as the project manager for a new project to design a new battery to operate electric cars. How would you use the 4-D Algorithm for New Product Design to assist you as the project manager and designer or engineer?

BIBLIOGRAPHY

Brown, T., *Change By Design, How Design Thinking Transforms Organizations and Inspires Innovation*, Harper Business, 2019

UK Design Council Website, *What is the Framework for Innovation? Design Council's Evolved Double Diamond*, UK Design Council Website, 2022

Winchester, W. *Inclusive and Consequential by Design: "Futurefying" New Product Development (NPD) Through Vision Concepting*, 2019. INCOSE INSIGHT, Volume 22, Issue 3, Pages 49–51, October 2019

Part 2

---

*How Does Organizational
and Individual Readiness
Support Inclusivity into
the Design Process?*

6 Enterprise (Organizational) Pillars for Engagement Success

6.1 ORGANIZATIONAL COMMITMENT TO INCLUSION

Engineers may choose to use any design process, such as the *Framework for Innovation* or the 4-D Algorithm for New Produce Design, but no matter what process is employed, the organization itself is the foundation by which inclusion and diversity are incorporated into all work processes. As depicted in both the *Framework for Innovation* and the 4-D Algorithm for New Produce Design, the role of leadership and organizational dynamics are highlighted. The organization and its members, through implementation of policies, procedures, and decision-making establish what is considered acceptable behavior and the manner by which work processes meet expectations. **Inclusive organizations have a diverse workforce that value the opinions of all members and are able to work together sharing ideas to accomplish the goals of the mission.** Leaders in inclusive organizations generally focus on recruiting and retaining a diverse staff that reflects characteristics of the neighboring communities (race, ethnicity, culture, etc.) where business is being conducted. The management in these organizations possess exceptional leadership skills that are evident to, and embraced by, organizational members. In addition, the leaders are inspirational and are able to inspire others to follow and ignite their creative abilities to improve processes and systems and develop new processes, products, technologies, and systems that will further enhance the goals and mission of the company. In order for organizations to cultivate an inclusive culture they must

- Exert effort supported by developing and implementing policies that can facilitate the desired culture
- Engage in actions that model the appropriate practices as they go about their daily activities
- Be championed by a leadership team that is empathetic and has the ability to set the example for others to follow
- Be open to new ideas that do not typically follow what has always been done

The United States (US) is an ever evolving multiethnic and multicultural country. This cultural diversity makes the US an interesting and exciting place to explore, live, and work. It also renders the workplace environment challenging when it comes to integrating diversity into a work environment in a manner that provides enhancement to the workplace culture. A positive workplace culture is recognized as an attribute in enhancing technology discovery and the quality of products produced

DOI: 10.1201/9780367854720-8

by an organization. The cultural makeup of organizations within the US is quickly resembling populations from around communities and across the world. Many from other countries visit the US annually for vacationing or visiting relatives and friends that currently live in the US, while others enter the US to continue their education, to begin their careers in the US, or migrate to the US for a better lifestyle. This influx of people from other countries contributes significantly to the cultural diversity that must be included as a part of the culture of the US and accounted for as a part of the workplace environment if success and inclusion in organizations is to be achieved.

Workforce diversity (ethnicity, nationality, cultural background, religion, gender, age, way of thinking, etc.) can become a competitive advantage when problem-solving, developing innovative product ideas, and targeting marketing initiative (Pless and Maak, 2004). This advantage is only realized if the right processes, systems, culture, and leadership team is in place and are cohesively functioning. Today corporate leaders realize that they must build a diverse organization and move into the realm of inclusion to further position the organization and company to reach its highest level of performance. To leverage workplace diversity, management and employees must be actively engaged in the process and provide support to develop the level of cohesion and relationships necessary to move an organization in the direction of collectively performing at a higher level. As such, employees and managers have important roles in developing and maintaining an inclusive organizational culture. **The integration of the role of leaders and organizational members together forms the basis of how an inclusive culture is formed and maintained**. Figure 6.1

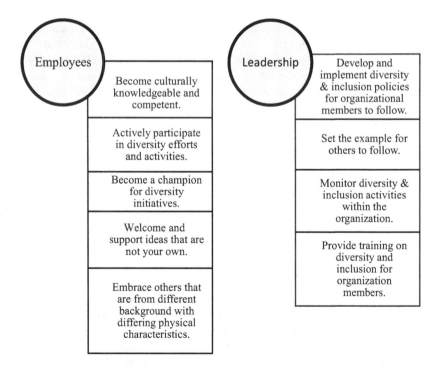

FIGURE 6.1 Inclusive Culture Roles.

lists some of the important roles that employees and managers must exhibit to move the needle in the direction of developing, embracing, and maintaining an inclusive workplace culture.

An organization must be able to produce the products or deliver the services that meet the needs of their customers to be viewed as and become successful. A diversified organization can lead to unique ways of solving problems and the creation of products that can be utilized by a diversified populace.

6.2   THE FUTURE-FORWARD TECHNOLOGY ENTERPRISE

Many people believe that the industry of the future will primarily consist of automation using robots or moving parts in filling orders and delivering products to customers. Technological advances continue at a rapid pace and is more prominent and futuristic today than in previous years. One may recall viewing a sci-fi movie decades ago and witnessing the actors or actresses moving from place to place on an object that resembles what we know today as a hoverboard or a flyboard that is capable of rendering a person airborne for some period of time. Futuristic designers are keenly aware of the history of product design and development. These designers use considerations of current information and technologies and that of the past in designing technology and products for the future.

The ability to keep up with the many technological advances and getting in front of these advances is challenging for many businesses and their workers. In a sense, in order to keep up and surge ahead of technology advancement an individual and corporation must take a futuristic view of technology development and implementation and identify what role they would like to play in that advancement. **A futuristic view requires an organization to empower their workers to be able to recognize and evaluate current technologies and anticipate the needed changes for future use.** It also requires a focus on research and the development of new technologies and products. In part engineers and designers must be allowed to imagine and explore what may seem impossible today. For example, during early years while watching the show *The Jetsons*, it was common for the cast of this popular cartoon show to move around from place to place in a car that travelled through the air. Today flying cars are being manufactured and tested for consumer usage in the future.

Looking through a futuristic design lens is not easy as it takes having strategic leaders and professionals that are willing to think outside of the box and explore concepts at times that are complex and time-consuming to evaluate and develop. At times these concepts may even seem unrealistic on the surface; however, once explored the possibilities and practicality becomes clear. What does it really mean to futuristically approach technology?

The first challenge begins with employing a strategic forward-thinking leadership team that not only looks at product and process needs of today but will also have a focus on what will be needed in the future. A futuristic view of business operations will serve a company well in developing products and services that will meet the needs of their customers. In order for employees to be successful taking a futuristic approach, they must be onboard and know that they have the support of leadership in developing their ability to explore, think, and perform creatively.

6.3   CREATING AND NURTURING A CULTURE OF DIVERSITY AND INCLUSION WITHIN THE TECHNOLOGY ENTERPRISE

There are very few world events or activities that are not driven or influenced by various aspects of technology. The global economy has been rapidly changing, and technology development and advances are playing a key role in rendering these changes successful with the development of new products, concepts, practices, processes, and technologies. Employment in high technology industries continues to increase adding benefits to the national economy (Alston, 2014). Engineers and organization leaders do, and will continue to, play a pivotal role in the rapid changes seen today and into the future.

One of the most important roles of a leader is to drive the culture of an organization. The culture of the organization has far-reaching implications in terms of

* The way work is performed
* The relationship among organizational members
* The relationship and services provided to customers
* Integration and inclusion of workers within an organization
* Development and implementation of procedures and policies

(Allen, Alston, and Dekerchove, 2019)

The organizational culture that must be developed, nurtured, and maintained must be one that has the ability to excite and encourage, build team cohesion, lasting relationships, enable personal growth among people, and provide the tools that people can use to ensure that they can achieve their best performance. When we refer to an organizational culture, we are not just referring to one organizational characteristic, the overall organizational culture is a significant driver of an individual's viewpoint and expectations, and these viewpoints drive organizational behaviors (Gordon, 2017). The shared values and beliefs that exist within an organizational culture represent the significant variables that guide acceptable behaviors.

The organization's culture has a prominent impact on behaviors as well as impacts many aspects of the organizations' life to include how people are treated, implementation of processes and procedures, and the distribution of reward systems. Therefore, significant influence is placed on organizations by their culture (Alston, 2014). These influences can be positive or negative depending on the viability of the culture, the mindset and engagement of the organizational members, and the influence of the leadership team.

According to Pless and Maak (2004), an organizational culture of inclusion needs to be established in order for the potential of a diverse workforce to be released. A culture of inclusion is important in cultivating integration between workers from diverse backgrounds with diverse thoughts. **An inclusive organizational culture builds on the normative grounds that recognize the differences, as well as the similarities of individuals.** In addition, an organizational culture of inclusion allows people from multiple backgrounds and ways of thinking to work together in an efficient manner while achieving the objectives of the organizations; all voices are heard and respected while demonstrating value for the various perspectives and approaches voiced.

6.4 CULTURAL ENABLERS FOR SUCCESS

Organizational cultures have been studied for decades, and as such, much is known about the value of culture in an organizational setting. For example, Sathe (1983) advocates that substantial influence can be placed on an organization because cultural elements can guide behaviors. Harrison and Stokes (1992) states that not only does culture influence members of an organization, it can also impact many aspects of organizational life that includes promotion and reward systems and how people are treated. The documented research on culture states that culture has wide-reaching implications for organizations from the way work is performed, the relationship among employees and with customers, in developing and implementing policies and procedures, and acceptance and integration of a diverse workforce.

It has been said many times that the United States is recognized as a nation of immigrants. With this recognition, the reality becomes that as a nation of immigrants, the culture of an organization becomes more complex and difficult to facilitate integration into a process which enables development of new designs and products. **When reviewing the way culture is defined broadly by most theorists and practitioners, there is agreement that culture encompasses the practices, values, beliefs, behaviors, and patterns that are evident in the organization that forms the identity of the organization.** Therefore, when developing or evaluating workplace cultures, these attributes must be included in the review. Developing an organizational culture that can stimulate and sustain business growth is incredibly challenging, primarily due to the diversity of cultures, different thought processes, the different rituals people bring to organizations, and varying subcultures that underline the overall culture of that organization (Allen, Alston, and Dekerchove, 2019). Therefore, it becomes necessary for an organization to infuse characteristics that can enable the type of culture desired. These characteristics are referred to as culture enablers.

Cultural enablers are defined as those characteristics within an organization necessary to encourage, support, and sustain diversity in thoughts and actions of members as they embark upon the design, development, and implementation of technology that will be introduced into a culture of diverse people. The most important aspect of enabling culture is the leadership team coupled with ensuring that the employees supporting the organization are engaged. It is necessary for everyone in the organization to utilize a set of policies and procedures that support cohesive operation of the many organizational systems. This relationship is shown in Figure 6.2 with further discussions in sections 6.41–6.6.3.

6.4.1 PROFICIENT LEADERSHIP

Having the right leadership team in place is extremely important to a company in developing and implementing its business strategies, discovering new technologies, and achieving the goals and mission of the company. Without the right leadership team, a business can quickly become ineffective and unprofitable and will eventually cease to exist. Leaders are not always selected with consideration of how to enhance organizational effectiveness and improve services and product delivery. In many organizations, leaders are selected because they were good individual contributors in their areas of

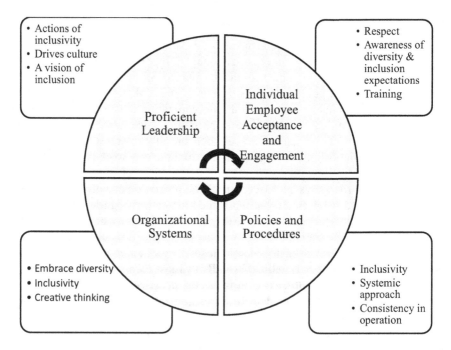

FIGURE 6.2 Cultural Enabler Matrix.

expertise; they manage a project successfully; they are good with delivering what their customer is seeking, and so forth. These reasons are not bad reasons for selecting a leader; however, there are many more qualities that must be considered if the leader is seeking success and has the desire to be viewed as a competent leader.

When selecting leaders, there are several key attributes that should be considered to begin with; and yes, there may be a need for additional training and mentoring for the right candidate to successfully perform their leadership roles. Attributes to be considered when selecting managers and leaders should include the following:

- Evaluate the position and identify key leadership attributes that are needed to achieve the goals of the organization
- Evaluate problem-solving skills
- Evaluate decision-making abilities and the ability to integrate and assimilate feedback from others in decisions
- Evaluate communication skills both verbal and non-verbal
- Does the candidate have the ability to motivate and inspire others to follow?
- Does the candidate have good interpersonal skills?
- Evaluate customer service skills and philosophy
- Evaluate leadership style and appropriateness for the organization
- Evaluate trustworthiness characteristics

The above list is a good start; however, based on the goals and mission of the organization, additional characteristics will be important.

6.4.2 INDIVIDUAL EMPLOYEE ACCEPTANCE AND ENGAGEMENT

When employees share the vision of the organization, the management of the company flourishes across the board, from support and adherence to policies, respect for the leadership team and each other, procedures to increase productivity, and providing quality customer care. Employee engagement stems from the emotional connection one feels toward their management, other employees, and the company and is primarily relationship-based. This connection has great influence on the behavior and the effort an employee exerts when performing work, with interactions with leaders and other employees. When an employee is engaged, they are generally passionate about their jobs, willing to go the extra mile to meet a goal, engage in creative thinking, and demonstrate commitment to their organizations. According to Alston (2014), employee engagement has many benefits for organizations, including

- Increased productivity
- Increased facilitation of employee growth
- Empowered employees who participate in the business
- Increased trust
- Facilitated employee retention
- Increased creative and strategic thinking
- Better decision-making

In addition, employee engagement is key in implementing the 4-D Algorithm for New Product Design discussed in Chapter 5, effective usage of Design Lens and vision concepting concepts discussed in Chapter 4, and in the design and development of technologies.

Not only must employees be engaged in their organizations, but they must also believe and feel like they are valued and included as an integral entity in helping achieve goals, objectives, and the mission. **An inclusive work culture is necessary to tap into the creative thinking and ability of workers to enable them to have a sense of belonging and responsibility for the success of the organization.** Inclusion and acceptance of diversity in organization will not be successful if the workers are not in support of and are engaged in activities that will render the viewpoints of organizational members to support various thoughts and ways of accomplishing work. Engagement and inclusivity are integrated in facilitating organizational success.

6.4.3 POLICIES AND PROCEDURES

Policies are developed by organizations to influence behavior, set boundaries, determine how decisions are made, and the types of actions that should be taken in various situations. Procedures outline specific methods employed that express the policies and the way business is carried out on a daily basis. These policies and procedures also set the stage for how organizational members are to act, react to the various situations that may occur, the expectations for the way they interact with each other, and are instrumental in assisting an organization is achieving its goals. Some of the benefits of having policies and procedures in place include

- Clarifies the expectations of organizational members
- Provides an understanding for employees on their job function removing trial and error when performing work
- Provides a mechanism of control for an array of topics and practices
- Establishes and documents roles, responsibilities, and accountability

When developing policies and procedures, consideration must be given to the needs of the business and what outcome the policy or procedure is expected to resolve, regulate, or achieve. Without policies and procedures, members are working inconsistently, disjointly, and have no mechanism in place to assist them in strategically functioning to achieve the mission set before them.

6.4.4 ORGANIZATIONAL SYSTEMS AND FUNCTIONALITY

Organizational systems play an important role in the function of any organization in terms of efficiency, focus, and ease of accomplishing work. These systems encompass the structure of how information flows between the various levels in a company and how the organization is structured, organized, and arranged. Many organizations are organized into one of four systems that are known as matrixed, functional, flat (horizontal), or divisional. **A well-defined organizational system provides to all employees' information on what is expected of them and who will directly evaluate their performance**. A point to note is that organizational systems should be designed specifically for a company with their mission and goals in mind. Also, organizational systems can change over time depending on the changes in business strategy and mission of a company. An organization will typically use one or a combination of four systems. These systems are discussed in more detail below.

6.4.4.1 Functional Organizational Systems

A company that is functionally organized is considered to have a traditional hierarchical system. **Team members are grouped and organized based of their area of expertise in a functional organizational system**. Generally, the manager or supervisor for the organization has similar or the same type of expertise; and therefore understands the knowledge distribution of the staff and is able to effectively strategize and utilize the skills of workers to ensure that the mission and goals of the organization are realized. This type of structure may be optimal for encouraging inclusion of workers, different cultures, voicing differing thoughts since workers share a common characteristic based on their chosen field, knowledge, and expertise. However, complexities may come into play when the various functions have the need to work with other functions with differing knowledge levels and experience with a subculture. These structures are arranged in specialized divisions, such as finance, engineering, marketing, human resources, infrastructure, or information technology. The reporting line in functional structures are clear as each function generally reports to a single senior manager with similar skills. In addition, employees tend to become specialized, and communication can become an issue when the need arises to work with other functions. Organizing based on functions may exhibit both disadvantages as well as advantages. Some advantages include

- There is no duplication of work efforts since employees have specific job responsibilities within the function.
- Employees are grouped by their knowledge and skills, which facilitates the achievement of a higher level of performance.
- The highest degree of performance and collaboration occurs because employees are grouped by their knowledge, skills, and expertise.
- Career path and growth opportunities are clear.
- Communication is clear and fluent with the function.
- Work accountability resulting from fixed roles and responsibilities
- Easier to foster and achieve a sense of inclusion and comradery among workers

Some disadvantages include

- Employees may become bored and myopic and may lose enthusiasm and creativity because of the repetitive work activities.
- Highly skilled workers are generally more expensive; therefore, payroll (labor costs) may be higher.
- Communication may become stovepiped within the function and poor with other departments.
- The functional structure is generally rigid, which makes it difficult to implement and for members to adapt to changes.
- Decision-making can be slow due to the bureaucratic hierarchy.
- Decisions are generally made at the top by the functional manager and flowed down to the rest of the organization.
- Functional departments become focused on the goal of their department and not on the goals of the entire organization as a whole.
- For large functional organizations, each department can have their own culture and management style and may act like small companies within each functional group.

6.4.4.2 Divisional Organizational System

A divisional system is a type or organization designed that groups together employees that are working on a specific project, product line, or marketing service. **In a divisional system, employees are very familiar with their team's work, although they lose familiarity with the work that is performed by other teams.**

Some advantages include

- Allows the team to focus on one project or product. Focusing on one project or product provides an opportunity for the selection and use of the appropriate design lens to facilitate design and development of inclusive products.
- The leadership team also focuses on one project or product line.
- Ease of building one culture and minimizing subcultures
- Expedited decision-making ability
- Greater level of accountability for the projects or product lines of responsibility
- Easier to respond to changes in internal or external environments

Some disadvantages include

- Isolation of workers in differing divisions, limiting knowledge of the entire corporate enterprise
- In many cases employees are unable to work efficiently across divisions.
- Stovepiped exposure to other products or projects that are a part of the company's goals and mission
- Potential increased costs for a company because of duplication of resources as each division will have some of the same resources performing similar tasks

6.4.4.3 Matrixed Organizational Systems

Matrix organizational systems combine the functional and divisional organizational structure reporting systems. **The matrix organizational structure can be complex and requires a vast amount of planning and an effective means of communication across the organization.** Many large organizations and organizations that have a variable need for differing skill sets at differing time, tend to utilize the matrixed structure to guide efficient use of employees and specific skill sets. For example, in research and development organizations, some resources are aligned functionally, and many are matrixed out to a project or a division to serve as team members to accomplish team goals. Regardless of the organizational structure selected, there are distinct advantages and disadvantages that must be considered. The same is especially true when deciding to organize along matrixed lines.

Some advantages include

- Clear communication of project objectives
- More efficient use of resources
- Team structure encourages relationship building.

Some disadvantages include

- Employees generally have two bosses, and as a result, the worker can feel overwhelmed with differing priorities.
- Increased organizational complexity in reporting lines
- Potential increase in overhead cost due to the need for additional management
- Some employees may feel overworked when supporting several projects.

6.4.4.4 Flat (Horizontal) Organizational Systems

Flat organizational structures remove the hierarchal structure and allows employees more autonomy and creativity in getting work accomplished. **A flat organizational structure encourages employee engagement, innovation, and creativity because the bureaucratic processes that come along with a traditional hierarchical system are not pres**ent.

Advantages include

- Employees more than likely will be trained to perform more than one task.
- Increased communication

Disadvantages include

- Limited promotional opportunities and personal growth for employees
- Difficult to maintain as an organization grows
- Employees may become overwhelmed when performing too many tasks.

6.5 APPLIED LEARNING

1. Explain the contribution that diversity and inclusion have to an organization and its effectiveness.
2. How does the cultural makeup of the United States impact workplaces and the ability to develop new innovative technologies and products?
3. What role do leadership and employees play in workplace diversity and inclusion as it applies to a futuristic look at technology and product development?
4. What are some of the most important responsibilities of a leader in his role of leading an organization?
5. What is the importance of establishing the appropriate work culture?
   a. What role does culture play in effectiveness of an organization?
   b. How does leadership involvement facilitate culture development?
   Note: include the cultural enablers in your discussion
6. Discuss the function and role of organizational systems.
7. List and discuss the different types of applied learning concept and their impact on the work environment.

6.5.1 CASE STUDY

The president of a start-up technology company is in the beginning stages of designing his organizational system to support the business proposal developed and supported by his investors. As a start-up venture his staffing level will be around fifty employees initially with plans to expand to approximately two hundred workers within the year. The mission of the company is to develop and manufacture small propriety parts for a military weapon system.

1. Select and design an organizational system to support the organization described in the case study.
   a. Discuss why you selected the system being recommended.
   b. List advantages and disadvantages that may exist that will impact diversity and inclusion as it relates to technology development.

Additional questions:

1. Describe each organizational system and describe how each system can enhance or stifle new product development.
2. What are the benefits of employee engagement?
3. What are cultural enablers?
4. What is the role of management in enabling culture?
5. How does proficient leadership enable new product development and futuristic thinking?

BIBLIOGRAPHY

Allen Patricia, Alston Frances and Dekerchove Emily, *Peak Performance How to Achieve and Sustain Operations Management,* CRC Press Taylor and Francis Group, 2019.

Alston Frances, *Culture and Trust in Technology-Driven Organizations*, CRC Press Taylor & Francis Group, 2014.

Gordon Jon, *The Power of Positive Leadership*, John Wiley & Sons, New Jersey, 2017

Harrison, R., and Stokes, H., "Diagnosing Organizational Culture," Jossey-Bass/Pfeiffer 1992.

Pless Nicola M. and Maak Thomas, Building an Inclusive Diversity Culture: Principles, Processes and Practice, *Journal of Business Ethics*, Vol. 54, No. 2 (2004), pp. 129–147.

Sathe, V., "Implications of Corporate Culture: A Manager's Guide to Action," Organizational Dynamics, Autumn 1983.

Sibley Jim and Ostafichuk Peter, *Getting Started with Team-Based Learning*, Stylus Publishing LLC, Sterling Virginia, 2014.

7 Organizational Readiness and Design
The Engineering Management Imperative

7.1 INTRODUCTION

How does a leader know when an organization is designed to cope with current and future changes and challenges? This is not always an easy question to answer on the surface for most leaders because of all of the various parts of an organization that must be considered in developing an answer to the question. If an organization is meeting its goals, then usually it is assumed that the organizational design, along with its internal systems, are meeting the needs of the company and mission; leaders are prepared for whatever challenges they may face. The assumption is that the organization is on track to achieve its sustainability goals. However, this is not always true because the organization may not be functioning efficiently, and their goals may be marginally met. **The leadership team is key in organizational design, implementation, and maintenance, as well as to ensure that the leadership team are of the caliber that is needed to move the organization into the future.** There are twenty-one imperatives that are discussed in this chapter that are paramount in ensuring that the leadership team is ready to deal with challenges, develop inclusive designs and technologies, engage employees, develop, nurture culture and a lot more.

7.2 THE ENGINEERING MANAGEMENT IMPERATIVE (EMI)

Engineering management must be effective in bringing together the technological problem-solving ability of engineers and designers into operational aspects of the business enterprise. Management must also be capable of balancing the administrative, business planning, monitoring, and controlling activities in order to oversee the operational performance of engineering and technologically-driven business enterprises. There are at least twenty-one imperatives that engineering managers should be proficient in to optimize business outcomes and function in the peak performance realm. These imperatives listed in Table 7.1 have significant impacts on the people in the organization, as well as the entire business enterprise.

DOI: 10.1201/9780367854720-9

TABLE 7.1

21 Engineering Management Imperatives to Optimize Business Outcomes

1. Develop business strategies	2. Implement business strategies	3. Plan, lead, organize
4. Risk management	5. Manage knowledge	6. Facilitate collaboration
7. Change management	8. Creating value	9. Optimize resources
10. Act strategically	11. Improve performance	12. Coach and mentor
13. Establish technical goals	14. Staff development	15. Culture development and maintenance
16. Provide safety and security	17. Facilitate communication and collaboration	18. Encourage creativity and innovation
19. Assure continuous improvement	20. Inspire loyalty and trust	21. Assure sustainability

7.2.1 DEVELOP BUSINESS STRATEGIES

Charting the course for an organization resides with the senior leadership and their team. A business plan provides direction that can lead to a more efficient and profitable organization. An effective business strategy provides the opportunity for an organization to forecast business needs current and into the future. In addition, a strategy will allow a company to engage in innovative technological design and product development to manage the demands of a changing marketplace. **A business strategy is a documented plan detailing the mechanisms used by an organization to achieve their goals**. A well-defined plan gives direction, drives the decision-making process, creates a measure of and for success, and increases adaptability of changing market conditions. Your business strategy must not only chart the path of the organization leading toward goal achievement, but it must also provide a path to sustainability.

7.2.2 IMPLEMENT BUSINESS STRATEGIES

Developing a business strategy is not enough if the desire is to promote and sustain an organization and prepare for future ventures. The business strategy developed must be implemented and not be placed on the shelf to satisfy auditors or the company board of directors' requirements. **The second stage of strategic management is implementation where tactics, outlined in the business strategy plan, are transformed into action which leads to performance improvements.** Strategy implementation is the most stringent and challenging aspect of a strategic management process. This stage of the process requires the use of the organizational resources to achieve success. The basic tenets or tactics of implementation, should include the development of objectives for implementation that are complementary to the strategy, development of policies and procedures that will be used to execute the strategy, performance of tasks and activities needed to achieve the strategic vision, allocation of needed resources, and leading, controlling, and measuring performance of various activities.

7.2.3 PLAN, LEAD, AND ORGANIZE

Fundamental roles assigned to leaders and managers are to plan and organize functions and activities that take place in an organization. Planning involves setting goals and determining actions necessary to achieve those goals. Some advantages of planning include the reduction in wasteful activities and resources, reduction in the risk of uncertainty, provides direction, promotes creativity and innovation, and enhances decision-making. Effective planning requires that managers be knowledgeable of current conditions facing their organization and forecast potential future conditions. **Organizing by management involves developing job responsibilities, organizational structures, and allocating resources to ensure that goals of the organization can be achieved.** Leadership, as opposed to commanding, requires that the manager has the respect and trust needed to insert influence that will inspire actions taken by others and the desire to follow. Managers who are skilled in planning and leading organizations have the basic skills needed to develop and implement a strategy that will chart the course and propel an organization in the right direction toward peak performance.

7.2.4 RISK MANAGEMENT

Many leaders struggle with enterprise risk management. **The process of risk management involves the identification, evaluation, and prioritization of risks.** Once risks have been prioritized or ranked, a coordinated effort to eliminate, control, minimize, and monitor the probability of occurrence is required. The risks that a company may face could result from a wide variety of sources, such as financial uncertainty, accidents, legal liabilities, and information technology threats. An effective risk management plan allows businesses to prepare for unexpected events by attempting to minimizing risks and reduce harmful events. Commercial software is available to help organizations manage risks. There should be a recognition that all risks can never fully be avoided or mitigated because of limitations such as the cost associated with mitigation and the practicality of identifying a solution for mitigation. Once the plan and risk register has been completed, there should be a process to validate the information. The requirement for an annual review of the risk register should be incorporated into the plan to verify and validate whether changes are warranted.

Another avenue to use in the management of risk is to effectively use design lens in the design and product development process. Different design lens cannot only help target, design, and develop products to serve a wide array of customers, but it can also help design out risks associated with the product to human interface. One cannot help but question whether the risks associated with the products below could have been minimized or prevented had design lens and vision concepting been embedded in the approach to design and product development.

- The case of artificial intelligence—It has been reported that the accuracy rate of this technology is higher for light-skinned people and lower for dark-skinned people.

- The case of Fit Bit—It has been reported that the product does not work as effectively for people of color, those who are suffering from obesity, or those with tattoos.

The risk associated with flawed designed product cannot only cost money, it can also damage the company brand and reputation.

7.2.5 MANAGING KNOWLEDGE

Knowledge management is the process of capturing, dispensing, and controlling the knowledge and information that is inherent within organizational members. **The goal of a comprehensive knowledge management process is to improve performance, enhance the competitiveness of an organization, continuous improvements of organizational processes and practices, and enhances the sustainability.** A significant part of knowledge management involves knowledge retention which involves capturing knowledge so that it can be used later. In order to maintain or control the loss of knowledge in environments where there is a high turnover of workers, companies are using tools to capture information from workers before they exit the company. The process used to capture knowledge can range from an interview of the individual using a scribe to capture information disclosed to using a software to electronically capture and retain the information.

7.2.6 FACILITATING COMMUNICATION AND COLLABORATION

Collaboration and communication are at the top of the list when it comes to engineering and design managers being able to gain access to technology discovery and improvement, service improvement, and new product development. Collaboration brings people together allowing them to share ideas and work together to achieve a common goal. **Collaboration benefits an organization by increasing teamwork, facilitating creative thinking, brainstorming of ideas and solutions, accepting and valuing the diversity of thoughts, employee engagement, knowledge and skills development, and enhanced problem-solving.** In some organizations collaboration workspaces have been developed to signal to workers the expectations for, and embracement of, workers getting together formally or casually to discuss workplace topics and issues. Effective communication forms the basis for effective collaboration, and the yielded benefits are improved processes. Open and honest communication that flows throughout the organization is an asset for any business environment for a host of reasons. Some of these reasons include assistance in avoiding and mitigating conflicts, increasing employee engagement and interactions, improving productivity, improving relationships among workers, management, and clients, and building and sustaining trust among organizational members. In order to move an organization forward and to arise as a significant player in a technology driven environment, engineering and design managers must set the expectations, embrace collaboration, and encourage open honest communication between employees.

7.2.7 CHANGE MANAGEMENT

Change is the one sure thing that will continue to happen over time. Many of these changes are necessary and orchestrated, and there are times when change may be swiftly occurring. **Change management is the process used to manage change within an organization, whether the change is due to increasing or decreasing the number of employees, eliminating a product line, or management change within the organization.** Change in organizations is usually upsetting and creates fear within workers when they are unprepared to deal with changes and when the culture is not one that facilitates or enables change. Because of the fear of the unknown, workers will automatically begin to resist the change and close themselves off to the potential need and benefit of the change. The engineering or design manager is responsible for recognizing when changes are needed to support the health and productivity of the organizations needed to keep up with competing in an ever-changing technological global environment. Assisting employees with managing and embracing change is a role of the leadership team. The engineering or design manager is responsible for bringing his/her managers along and assisting them to function as change agents to assist workers in accepting, and yes, embracing change as it occurs in the organization. Some actions management should consider in facilitating change include the following:

- Develop and implement a change management strategy that includes a comprehensive communication plan that is inclusive and proactive
- Ensure the reason for the change is clearly communicated as often as needed so that employees can gain an understanding of the importance and need. Open and honest communication is a necessity.
- Seek buy-in and feedback from workers especially those who will be of greatest impact
- Identify and enlist workers that support the change that can help sell the change to their colleagues
- Solicit feedback and be responsive to questions and information received

The ability of an organization to successfully implement change requires a culture that is supportive of, and able to, adapt to change. Establishing a culture that can tolerate change begins with management and an open team dynamic.

7.2.8 CREATING VALUE

Creating value is the substance on which to build a business and manage an organization. It is management's responsibility to ensure that business practices increase the value of services and goods they provide, which will in turn increase value to their customers. Before value can be delivered, a determination must be made as to what represents value. It is also management's responsibility to create value for the organization and its members. Delivering value to the customer is essential for sustaining and increasing profitability. In order to create value, there must be an understanding of what represents value for customers and the ability of company leaders to communicate the value of their design and products. In addition, the value created to support customers and the organization must be capable of withstanding the test of time. Creating value begins during the design stage and with the deployment of design lenses which promote development of products that are safe and can be effectively used by customers.

7.2.9 OPTIMIZE RESOURCES

The manager is charged with the responsibility of supporting organizational members and connecting the right workers with the needed resources. Organizational resources can be grouped into four categories that include human, capital, monetary, and materials. **The goals of the organization will not be achieved without employees having access to the needed resources, whether they are financial or associated with people, land, or equipment.** These resources encompass all assets that a company needs that are available for use in providing services or producing products. Management is charged with efficiently utilizing resources with minimal waste in order to render the organization productive and financially stable. Effective balance and usage of the various types of resources requires a skilled strategic leader who has the ability to inspire others to embrace the goals of the organization.

7.2.10 ACT STRATEGICALLY

Managers must act strategically when conducting business on behalf of an organization. In order to act strategically, they must be methodical thinkers so that their decisions and actions can guide the organization and its members into a profitable present and future. **Acting strategically means that the manager is capable and skilled in developing and implementing a strategy through execution of defined tactics that are supportive of the goals of the organization.** They are also skilled in using the resources of the organization in a manner to ensure that the organization achieves the tasks necessary to accomplish the goal efficiently. A strategic leader focuses on the health and viability of the entire enterprise. Their strategy should include a focus on the following, at a minimum:

- Knowledge management
- Attracting and retaining employees
- Providing customer value
- Keeping track of competitive market conditions
- Establishing objectives and tactics that will achieve the organizational goal
- Analyzing the competitiveness of their business environment and market and adjusting when needed

7.2.11 IMPROVE PERFORMANCE

Operational performance in organizations is of constant concern for company leaders. Business growth and sustainability is based on performance. **Improvement of performance takes on two facets, performance of workers and performance of the organization as a whole.** When considering performance of the organization, reference is also being made to systems, processes, and practices. Therefore, leadership must pay attention to a wide array of activities in order to continuously improve performance. To improve performance in organizations management must be able to.

- Engage and empower employees
- Identify and address roadblocks
- Develop and align meaningful metrics that are shared with employees
- Focus on the organizational strategy and revise as needed
- Maintain a well-trained staff
- Build and sustain trust
- Ensure that employees understand and support the business strategy
- Ensure adequate tools and work environment

Management must be able to recognize when an organization is not performing according to expectations, goals, and objectives. In such case, a performance improvement plan may be necessary to get the organization back on track. **The organizational performance improvement plan should take into consideration the human element, different design lens, as well as the impact of organizational systems, practices, policies, and procedures.** *Coaching and Mentoring*—Developing individuals and management can occur using various methods such as coaching and mentoring. These methods are commonly used as a more cost-effective means to develop people and build relationships. The two methods are performed by management on a daily basis as they either formally or informally perform their roles. **Employees observe management and at times emulate their actions, and some employees look to managers as a role model of who they would like to become as leaders in the organization**. Coaching and mentoring employees should be an integral part of each management's practices; however, it takes continuous focus and effort. The organizational benefits of management coaching and mentoring employees outweighs the time and effort invested. Benefits that can be achieved include building relationships with employees that lead to a sense of connection with the organization, further development of skills and knowledge, enforcement of the performance expectations being sought of employees, building trust between management and employees, and more.

7.2.12 ESTABLISH TECHNICAL GOALS

Goal setting is the prime responsibility of management. The engineering or design manager is charged with ensuring that the technical goals of the organization are set and incorporated into the strategic plan. Establishing organizational goals is one of the more challenging responsibilities of management. In setting goals, attention must be paid to establishing appropriate long-term and short-term goals. An important aspect of goal setting is that it can be used to motivate employees and keep them aligned on where the organization is heading. Important concepts to keep in mind include ensuring that

- Goals are clearly defined.
- Goals are communicated.
- The means for accomplishing the goal is determined.
- Appropriate metrics to measure and communicate results are developed.

Reasons for establishing goals may include providing direction and guidance on where the organization is headed, determining performance across the organization, and enhancing employee engagement and inclusion and can be used to forecast and allocate all types of resources.

7.2.13 STAFF DEVELOPMENT

Maintaining a skilled qualified staff, which ensures the goals of the organization are accomplished, is the responsibility of management. There are different means available to management to develop and maintain a highly skilled staff. Employee development must be included in the business strategy for the organization and funds allocated to support development activities. Development activities can be in the form of formal classroom training, mentoring, coaching, networking, on-the-job training, job shadowing, etc. When developing employees, consideration must be given to the employee and the role they play in the company, the type of development activities that will enhance the employee's performance, and the needs of the company. For organizations focusing on technology development, specific training may be needed to focus designers and engineers on thinking freely and exploring their creative abilities to develop products that can withstand the test of time. Specific training may include effective usage of design lens or how to develop products that are safe and inclusive. Regardless of which methods are employed, there is as associated cost either financial or time allocated to administer and participate in the development activity. Maintaining a skilled workforce can have advantages, such as retention of employees, productivity increase, creative thinking, improved problem-solving, and employee engagement.

7.2.14 CULTURE DEVELOPMENT AND MAINTENANCE

Many scholars recognize the culture of an organization as the basic foundation of how work is performed, how people interact with each other, and how customers are valued. **The organizational culture forms the basis for and enhances the ability of team members to build relationships and trust among management and their colleagues**. Just as important to recognize is that human performance is significantly dependent on culture because culture has an influence on people in terms of their ability to release creativity and innovative thinking that leads to increased productivity. The fundamental basis of design lens is built upon the unique creativity of each person. A company's culture can be a viable and sustainable competitive advantage because culture provides a bond that motivates organizational members. The shared values and beliefs that are present within a culture are important variables that guide behaviors. Because of the recognized benefits of culture, management must pay attention to, and engage in, activities that will build and sustain the appropriate culture while increasing productivity and sustainability of the organization.

7.2.15 PROVIDE SAFETY AND SECURITY

Safety and security are two of the most vital aspects of a business that prospective employees seek when deciding whether to work for a company. It is the responsibility of management to provide a safe and secure workplace for employees. According to the Occupational Safety and Health Administration, employers have a legal obligation to ensure a safe and healthy workplace for employees. Management must also ensure that appropriate measures are in place to ensure security such as having proper monitoring devices in place to monitor access to the work environment. Workplace safety and security are important because a feeling of security and safety when working can lead to a positive work culture and increased productivity. **A positive work culture that includes leadership demonstrating concern for employees is a significant factor in keeping employees feeling secure and motivated.** Safety and security are key elements that should be considered when design, developing, and deploying products and technologies for others to use.

7.2.16 FACILITATE COMMUNICATION AND COLLABORATION

Effective communication and collaboration are key to the success and long-term viability of a group, team, and organization. Communication and collaboration work hand in hand in getting work activities accomplished. An individual cannot collaborate if they are unable to effectively communicate. Collaboration must be effectively used by organizations especially those that are engaged in research and development and product or technology development. **The collaboration process involves having two or more individuals working together exchanging information and knowledge to achieve a goal.** Collaboration is also beneficial in building working relationships, enhancing employee engagement, and inclusion. To facilitate communication and collaboration, managers should

- Create a culture that supports collaboration and communication
- Provide communication and collaboration areas for employees
- Encourage employee engagement
- Encourage social interaction and positive dynamics within the group
- Use video conferencing
- Engagement in fluent, open, and honest communication

7.2.17 ENCOURAGE CREATIVITY AND INNOVATION

Creative and innovative employees are the types of employees that many industries seek to hire because of their ability to develop novel approaches and solutions. Creative thinking can lead to creative problem-solving and innovative thinking can lead to technology advancement. **In order for employees to feel safe in exercising their creativity, they must believe that they are in an environment where leadership is supportive and encourages employees to think outside of the box.** Outside of the box thinking, executed through brainstorming, is not an exact science, and mistakes can be made during the process; however, the benefits can be life-changing for a company in terms of productivity, sustainability, and the financial gains realized. Innovation within organizations requires a supportive and flexible culture that supports and nurtures discovery.

Attributes seen in an innovative and creative organizational culture include the following:

- Members are always thinking about how things can be done better.
- A focus on research and development and enhancing technology
- Employees are engaged and are encouraged to speak freely.
- Openness to, and acceptance of, the diversity of ideas
- Innovation and creativity are considered a core value and included in the company's business strategy.
- Trust is shared among organizational members.
- A willingness to share and support the vision

Key benefits of a creative and innovative organizational culture include

- Increased employee engagement
- Increased productivity across the organization
- Increased interaction relationship building among organizational members
- Increased team cohesion and inclusion
- Increased problem-solving efforts
- Increase in speed of technological innovation

7.2.18   Assure Continuous Improvement

Continuous improvement is the commitment of an organization to engage in ongoing activities that will provide improvement to their products, services, and the value delivered to both their internal and external customers. An integral part of the continuous improvement process is having a way to monitor progress and determine when changes are necessary, a system of checks and balances. As a part of the monitoring process, a good set of metrics is necessary to provide perspective in identifying and determining performance. It is management's responsibility to set the expectations for continuous improvement and ensure that the appropriate metrics and monitoring processes are in place and utilized in order to gauge and improve performance. Continuous improvement requires that the entire organization be involved and committed to improving across the enterprise. **When the organization is actively involved in reporting issues to management and assisting management in identifying solutions, continuous improvement then becomes a part of the business strategy and practice**. A continuous improvement culture requires the champion and support of top and line management to yield optimal results.

7.2.19   Inspire Loyalty and Trust

Trust and loyalty go hand in hand in organizations because without trust there can be no loyalty. Trust is becoming recognized as important to the success of corporate business strategies as management begins to understand and embrace the benefits of a trusting culture. **Trust is the central element in designing and maintaining a progressive organizational culture and is the foundation for success of**

high performing teams and organizations. Loyal employees are more devoted and attached to the organization and will actively engage in assisting the organization in achieving their goals. Some benefits of having employees who are loyal and trusting include the following:

- People are more willing to share information, admit to and learn from mistakes, and take on challenging assignments.
- Improved morale and productivity
- Creates an environment that encourages cooperation and permits employees to focus their attention on the tasks
- Enhances the quality of decision-making and the implementation of those decisions, while stimulating productivity
- Trust and loyalty together can be linked to increased performance levels, increased creativity, and critical thinking.
- Trust and loyalty help employees to be more committed to their work.
- Improved employee engagement

Many of the drivers for trust and loyalty formation are under the control of management. Therefore, management is the key to whether employees trust and become loyal participants of an organization. **In order to achieve trust and loyalty, managers should create a partnership with employees, be open and honest in communication, facilitate employee engagement in problem-solving, and provide a learning culture.** The Engineering Management Imperatives is key in proving a comprehensive view and approach in nurturing the people of the organization and the organization. Figure 7.1 presents a pictorial of where these imperatives may exert most influence is the overall health of an organization.

7.3 ASSESSING ORGANIZATIONAL READINESS

An organizational readiness assessment is a targeted, comprehensive measurement of the preparedness of an organization's ability to undergo change, and develop and implement new projects, processes, or technologies. The results gained from an organizational readiness assessment gives the leadership team an indication of whether the organization is prepared for and will be successful with their endeavor. This type of assessment may have more credibility when conducted by a third party who themselves are viewed as a being known for successfully executing assessments of processes and practices. Organizations continue to evolve and respond to changes resulting from environmental conditions as they attempt to gain market shares and create a culture that will ensure sustainability of the organization. As such, organizations must always maintain a state of readiness with the capability to adapt quickly to changes. To have the level of flexibility and fluidity that will render an organization successful when implementing innovative technology, developing new products or services, or entering a new market, it is incumbent upon management to understand the status and capability of their organization and employees. This can be accomplished through assessing your organization on some predetermined frequency. A comprehensive assessment program allows for evaluation of the entire

People
- Inspire loyalty and trust
- Encourage creativity and innovation
- Facilitate communication and collaboration
- Provide safety & security
- Culture development and maintenance
- Staff development
- Coach and mentor
- Improve performance
- Act strategically
- Change management
- Facilitate communication and collaboration

The Business Enterprise
- Assure sustainability
- Assure continuous improvement
- Provide safety and security
- Culture development and maintenance
- Establish technical goals
- Improve performance
- Act strategically
- Optimize resources
- Create value
- Manage knowledge
- Risk management
- Plan direct coordinate
- Implement business strategies
- Develop business strategies

FIGURE 7.1 21 EMIs Influence on Organizational Health.

organization to include evaluating risks and implementation of improvement activities, and the entire organization can be placed into three parts for the purpose of developing a comprehensive assessment strategy. The three parts of a comprehensive readiness assessment are listed in Figure 7.2. The three parts should continuously be assessed at a defined frequency to ensure that the leadership team is kept cognizant of the readiness of the organization.

7.3.1 PEOPLE

Humans are at the core of getting work accomplished in organizations. Therefore, a focus on the various aspects of what makes people successful helps to better

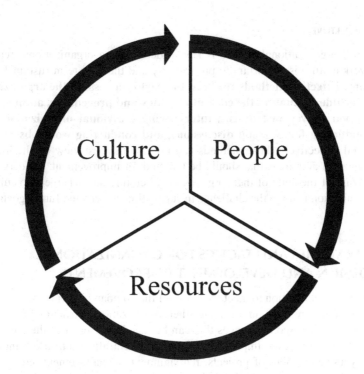

FIGURE 7.2 Readiness Assessment Components.

understand how they can contribute to the success of the organization. Keeping an awareness of the skill sets available for accomplishing current projects and the types of skills that will be needed for future projects is an aspect of managing an organization that a leader will want to always be aware of for strategic planning purposes. A skills assessment can be accomplished through conducting employee interviews using targeted questions to gauge knowledge and understanding in specified areas, performance reviews, and career planning discussions. Skills assessments not only keep leadership abreast of the talent that is readily available to management, but also offer an organization the ability to personalize training for employees and develop a strategy to meet the evolving needs of employees and the organization collectively.

7.3.2 RESOURCES

All resources used by an organization should be assessed to determine capability, operability, performance level, and life cycle status. These resources can include hardware, physical infrastructure, equipment, and IT infrastructure, processes, and systems. Because resources are different in needs and life span, function, and operability, it will be necessary to assess each resource carefully utilizing a process specific to that resource. When assessing some resources, it may be advisable to consult the manufacturers specifications to ensure complete understanding of the resource and performance expectations.

7.3.3 Culture

The overall organizational culture is key in determining organizational readiness as the work culture is important to productivity and the long-term sustainability of a company. **Effective methods that can be used in assessing the organizational culture include evaluating the content of policy and procedures, administering a perception survey instrument, interviewing individual organization members, conducting focus group discussions, and conducting work observations.** When used collectively these methods can provide the best view on the health of an organization and areas that should be targeted for improvement. Leaders should be cognizant of the state of their organizational culture and whether the culture of their organization is up to the challenge in supporting current and future goals of the company.

7.4 STRATEGIES AND TACTICS FOR ORGANIZATIONAL DESIGN AND DEVELOPMENT (IMPROVEMENT)

Designing an organization to meet the needs of the company is not a one-time event. Many companies are reluctant to change their organizational structure once in place because of the time and challenges that can be faced as a result of the change. For example, the change may impact the financial system alignment used, movement of employees, movement of projects, reassignment of management, etc. The current design of an organization should be evaluated on a predetermined frequency to ensure that it can continue to help the organization meet their goals when

- Development and implementation of a new business strategy could have wide-reaching impacts
- There is a change in business practices, technology, or market venture. These changes can be internal to the company or external.
- A recognition that the current organizational design is no longer effective in accomplishing goals and objectives

The design of the organization can be an effective process for organizing the manner by which an organization conducts business when aligned with the business strategy. Organizational design touches every aspect of organizational life including how policies and procedures are developed and implemented, reporting lines and responsibility, the decision-making process, communication channels, and the relationship between team members. When an organization is designed to complement the needs of the organization, some of the expected benefits include more effective and efficient decision-making, improved customer service and product quality, more efficient problem-solving, improved operational efficiency, engaged workforce, better prepared to face future changes and challenges. On the other hand, if the organizational design is not complementary of the needs and goals, significant problems and inefficiencies can occur to include inconsistent work quality, lack of coordination and communication across departments, waste of time and resources, and high staff turnover.

The most popular organizational systems that are being used by most companies at some levels are as follows:

- Functional—A functional system groups members of the organization based on their area of specialization and assigns leadership to the organization having similar knowledge and expertise. Workers are placed in units referred to as departments with a department manager who is held accountable for performance.
- Divisional—A divisional system is designed by splitting up an organization into semi-autonomous areas known most commonly as division. Each division consists of a complete set of functions that are essential to ensure that the division is successful in achieving goals.
- Matrixed—The matrix system combines two or more types of organizational systems. In this structure employees have more than one boss and may be expected to support more than one project at a time. Management responsibilities are often split between the functional manager and the project manager.
- Flat—A flat system is an organizational structure that does not have layers of management between the company's day-to-day staff members and senior manager. In these structures employees usually report to the president or the highest level of leadership instead to a lower-level management that reports to the senior manager.

No matter what system is selected for implementation, leadership should ensure that the system will serve the needs of the employees and the organization and enable work to be performed safely and efficiently, communication to flow, the needs of the customer served, and long-term sustainability.

7.5 IMPLEMENTATION OF BEST PRACTICES

A best practice is viewed as a protocol or process that has been accepted as being better than any other alternative used to address the same issue. Best practices are not difficult to identify when they occur because it is easier to recognize practices that are not commonly seen or ones that are highly effective in accomplishing a task or goal. Transferring best practices for one company or industry to another may be complicated if the practice is not accurately understood with the right tools in place to ensure success. **There are two primary methods that can be used to gain access to best practices: 1) one method includes looking within a company and identifying those practices or procedures that fits the category of a best practices, and 2) another method would be to look externally at other organizations practices and processes.**

When looking internally at the organization or company, obtaining a list of what is considered best practices across a company should be a business practice that is used by all management and work groups on a routine basis. Each organizational member should be expected to identify and share these practices with management as well as other groups within the company. These practices should be collected and documented in a comprehensive format and shared openly. Duplicating what is considered best practice can yield improvement and increase productivity in other areas of a company. Having a strategy to capitalize on what is considered a best practice can help a company become more competitive, become more efficient, improve the knowledge and skills of employees, reduce waste generated and disposed of, and improve quality and value for customers.

A widely used method of identifying and gaining access to best practices is through the process of benchmarking. **Benchmarking is the process of measuring performance of products, services, and processes from those organizations known to be leaders in one or more aspects of their operations against the organization desiring to improve their performance.** If conducted effectively, benchmarking can provide insight for organizations to compare their practices, processes, and procedures with similar organizations. The insight gained can help an organization identify areas, processes, practices, systems, or procedures that can be improved to facilitate productivity, increase value for their customers, and improve the financial bottom line.

There are times when implementation of best practices is challenging, and eventually fails because of issues resulting from transporting information, procedures, technologies, and practices. Therefore, it is incumbent upon management to ensure the accuracy of the information obtained regarding best practices and ensure that the best practices will fit within their current business strategy. Once best practices have been discovered and targeted for implementation in an organization, it is incumbent upon management to ensure the following:

- Demonstrated leadership support for the changes
- Selected best practices are appropriate for the organization
- The organization's culture can support implementation and sustainment of the best practices.
- The champion or liaison for implementation is prepared to lead the process with an effective implementation plan.
- Develop and implement a communication strategy to ensure that organizational members are kept informed and have an opportunity to ask questions and provide feedback
- Identify resources needed for implementation
- Ensure that the budget needed for full implementation is allocated
- Develop a schedule that is realistic and can be followed

7.6 APPLIED LEARNING

1. List the twenty-one engineering management imperatives. Select at least ten for discussion. Include in the discussion the following:
   a. A definition of the imperative
   b. How are the imperatives used by engineering managers in shaping culture and getting work accomplished?
2. If you had to add a 22nd imperative, what would it be?
   a. How would that imperative be used to assist engineering management in accomplishing the goals of the organizations?
3. List and discuss at least two of the imperatives that engineering managers can use to enhance people and business performance.

BIBLIOGRAPHY

Alston Frances, *Culture and Trust in Technology-Driven Organizations*, CRC Press Taylor & Francis Group, 2014

https://screenrant.com/apple-watch-fitbit-skin-tone-weight-affect-accuracy/

www.bostonmagazine.com/news/2018/02/23/artificial-intelligence-race-dark-skin-bias/

www.cnet.com/health/fitbits-might-not-track-your-heart-rate-right-if-you-are-a-person-of-color/#:~:text=The%20company%20said%20it%20designed%20its%20optical%20system,are%20cheaper%20than%20infrared%20lights%2C%20to%20take%20readings.

www.dailymail.co.uk/sciencetech/article-7325975/Fitness-trackers-fail-accurately-track-heart-rate-people-dark-skin.html

www.msn.com/en-us/news/crime/chicago-man-wrongly-imprisoned-because-of-artificial-intelligence/ar-AAZZzPR

www.theverge.com/2022/1/21/22893133/apple-fitbit-heart-rate-sensor-skin-tone-obesity

8 Preparing the Organization for Engagement
Engineering Management Enablers (Leading by Engaging)

8.1 INTRODUCTION

Just as managers work aggressively to transform and reinvent their companies through changes to processes, procedures, practices, and internal behaviors to improve performance, the same concerted effort and attention must be paid when seeking to increase engagement across the organization. Active engagement between team members within organizations has been receiving increased attention from business leaders because they recognize organizational success and productivity increases are often attributed to the engagement of their employees. Organizational engagement does not just happen; it takes vigorous efforts by leadership in developing policies and engaging in practices that are supported, embraced, and modelled by the leadership team. Many organizations who desire to develop an engaged workforce, struggle with achieving the level of engagement desired. Sometimes failure occurs due to the lack of modelling the desired behaviors by the leadership team. **Leaders must model the behaviors and actions that they seek from their employees in order for that behavior or action to be adopted and realized.** This principle is directly applicable to the designer or engineer who is leading or managing a team.

An organization or company's leadership team is the single most important enabler of organizational engagement, which facilitates new product development, culture development, and professional development of employees. They are responsible for organizational design, how employees interact with each other and the leadership team, and how the organization functions as a whole. The function of leadership can be complex and challenging and takes a skillful leader to fulfill the roles and responsibilities placed upon them. As such, it is important that organizational and company leaders possess the appropriate skills necessary to effectively function in their roles to ensure organizational success.

DOI: 10.1201/9780367854720-10

8.2 FOSTERING AND NURTURING THE CRITICAL AND REFLECTIVE PRACTITIONER WITHIN THE TECHNICAL ENTERPRISE

As the global economy continues to change at a rapid pace, technology plays a significant role, and technological advances continue to rapidly grow with the development of new products and reinventing or upgrading current products. With this acceleration of growth, leaders and employees must be equipped and adapt to compete and emerge as a sector leader. In order to be equipped to manage the changes, **organizations must develop and implement a strategy that is complementary to the vision and goals of the business.** Developing and nurturing professionals must be included in that strategy as they are at the forefront of technological advancement.

In this technological savvy world, there are few events or activities that are not influenced or driven by some type of technology. The global market for technology continues to grow while significantly contributing to the fast-paced economic growth seen in recent years. There are primarily three reasons why high technology companies play a significant role in the global economy. These reasons include

(1) The posture of high-technology firms is one of innovation. Companies that innovate are better positioned to gain and sustain market shares, create valuable new products, and become more efficient in operations, thereby improving their productivity.
(2) The research and development activities engaged in by high technology companies tend to be beneficial for other commercial sectors through generating new technologies, products, services, and processes.
(3) High-technology firms are actively engaged in the development of what most consumers believe to be value-added products. These products are also successful when introduced to foreign markets yielding higher compensation for a company (Alston, 2014).

Today more companies are struggling to attract and retain resources needed to ensure that business strategies are implemented. In fact, companies are becoming more creative by adding such attractors as free food to break areas, higher pay for specific skills, support for maintaining training and certifications, increasing the types of health care benefits, share in financial success, the ability to work remotely, and the list goes on. Often, much of the creativity deployed to attract and retain skilled employees are not deployed equally across the board; therefore, the success rate is not the same for every company. As such, the demand for skilled engineers and designers has created a favorable stance for the professional to choose the best opportunity for them, one that will allow them to contribute to the vision of a company, provide a learning experience, and advancement opportunities.

Before developing a strategy to attract critical staffing, the organization must first define what tasks or activities they consider to be critical for the company. **Critical activities include employee positions that are charged with performing tasks and activities that are necessary to support the goals of a company.** These positions are typically key to support development of products, technologies, and projects that are a part of the company's core competencies and are instrumental in the success of product lines of the business.

Conducting business in today's fast-paced economy, where there are a tremendous number of opportunities that talented workers are able to choose from, is adding notable stress and pressure on employers to engage in creatively identifying and employing activities, benefits, policies, and practices that will distinguish them from their competitors as the workplace of choice.

Once staffing has been secured and onboarded, another complex issue can be "care" and "feeding" of the employees to ensure that they continue to value being a part of the team in which they had just became a part, including maintaining their knowledge and skills. Technical professionals who are expected to assume an innovative posture, embrace diversity and inclusion of workers in their work environment, and function as collaborative team members need to be trained and mentored. A comprehensive training program should include the following at a minimum:

- Technical skills initial and follow up—regardless of the level of skills or previous training an employee may have had previously, there is always room for growth; therefore, additional training will be necessary.
- Soft skills—training in this area may include communication, presentation, and conflict resolution.
- On-boarding—training provided for new employees that provides exposure to company values and expectations, pertinent policies, and procedures of a company.
- Problem-solving—training provides knowledge on how to effectively solve problems as they occur to support the organization in improving efficiency and effective utilization of resources and reducing costs.
- Strategic Thinking—training assists an individual in developing the skills that will enhance their ability to anticipate and recognize issues and develop solutions while taking into consideration future ramifications. Strategic thinking skills can be helpful to designers and engineers when implementing vision concepting and design lenses concepts.
- Implicit Bias –training assists individuals in recognizing and gaining an understanding of the attitudes and stereotypes that unconsciously impact or influence an individual's actions and decisions.
- Diversity and Inclusion—training on the principle that everyone has something to contribute, and their differences can add value to the discussion and end product. Assembling diverse teams that include representation from a targeted population increases the ability for designers and engineers to design and implement products that can enhance the quality of life for a cross section of the population.

The example training types listed above are just a list of some of the educational opportunities that are key in developing and nurturing technical professionals including design engineers. These courses can help engineers and other professionals develop designs and produce new products, stimulate their innovative capability, and increase the level of productivity that is necessary to move a company in the top tier of their industry. In addition, investing in employees sends a message that the leadership team respects and values their contributions to the business. **Employees who believe that they are a valuable part of a team are willing to engage in the necessary activities to ensure the vision of the organization comes to fruition.**

8.3   TEAM DESIGN AND LEADERSHIP

Designing an effective team that is complementary of the needs of a product line, process, or strategy is dependent upon the vision of the leader for the organization and their prevailing knowledge, skills, and values. Team design is a direct responsibility of leadership, and when teams fail, leadership must take responsibility for the failure. Based upon experience, team design plays a significant role in whether a team is effective, marginal, or ineffective. Organizational design also plays a role in designing and supporting an effective team. Therefore, to have an effective team, leaders must take the time to design the appropriate organization and team in concert.

8.3.1   ORGANIZATIONAL DESIGN STRATEGIES AND TACTICS

Organizations are formed by bringing together groups of people for the expressed purpose of accomplishing goals and tasks that an individual cannot achieve in singularity. In order to achieve the goals, it is paramount to design the organizational system with careful consideration given to the type of system that will optimize performance across the entire enterprise. **An organizational system defines to members who is in charge of whom and the hierarchical roles and responsibilities of team members.** Prior to developing the organization, this framework must be strategically designed and implemented. This framework comes in the form of an organizational structure or chart.

A brief discussion of the different types of structures that can be employed within organizations will be introduced in Table 8.1. In some industries matrix systems are used extensively because they may be viewed as the most cost-effective system. However, in a matrix system, work performance may not be as efficient as with other systems because of the extensive sharing of resources. The divisional system should be considered and explored for new product designs as this system provide focus support for product development and implementation.

Before designing an organization or team consider the purpose, goal, and needed functionality to ensure optimal performance. Also, it is good to keep in mind that any organizational structure can be successful if the members of that organization want and desire it to be.

8.4   LEADERSHIP STYLES AND TECHNOLOGY DEVELOPMENT AND IMPLEMENTATION IMPACT

Throughout literature, many styles of leadership have been documented along with the expected characteristics of action that is expected based on the style they use in interacting with team members. Leaders are expected to be able to inspire and

TABLE 8.1

Organizational Systems and Deployment

Organizational System	Description	Optimal Deployment
Functional	This system organizes people based on their area of specialization and assigns leadership to the organization having expertise in the same area. Departments in this structure may be referred to as "silos" as they are focused on a particular function and tend to be isolated from other groups or functions.	Manufacturing organizations producing primarily the same type of goods
Divisional	This system groups employees together based upon the product line or market in which their work supports.	Optimal for companies with specific product lines that are different requiring different process and procedure to compete in the marketplace
Matrix	Combines the systems of a functional and a projectized system. This structure can prevent confusion for employees because they will generally report to more than one boss and may be assigned to more than one project at a time.	Optimal in organizations where critical skills are used on an intermittent or short-term basis to assist in providing expertise for a specific issue. Once solved the expertise is no longer needed.
Flat	This system lacks having multiple levels between employees and senior management.	Optimal for startup companies and companies that will remain small

motivate employees to accomplish activities and tasks to support the goals of the company or project. Motivation and inspiration are achieved through the use of the several styles of leadership. **A leadership style is the method used by a leader in providing direction, implementing a vision and plans, and motivating people**. Some leadership styles can impede technology development through actions, such as suppressing creativity among workers, discouraging and preventing the fostering of a diverse and inclusive culture within the organization, and the list goes on. Table 8.2 lists various leadership styles and their appropriateness in supporting technology development.

There is yet another style of leadership that was not included in the table. The style of leadership is referred as situational leadership (Hersey, 1984). It is conceivable that not one leadership style is appropriate for a leader in today's ever-changing environment as technology continues to rapidly evolve and worker engagement becomes the norm to achieve organizational success. A comparison of the various leadership styles with the situational leadership style attributes are shown in Table 8.3.

The leadership style employed is most likely to be based upon the situation to be addressed at that moment. Conceivably, a leader may employ a combination of the

TABLE 8.2

Technology Development Leadership Styles

Leadership Style	Description	Technology Organization Applicability
Autocratic (Absolute Leadership)	Autocratic leaders make decisions unilaterally and may not consider input or the opinions of others. These leaders make choices based solely upon their judgement or ideals, discourage creativity, tend to dictate how work is to be performed, and employees are not generally trusted to make decisions.	This type of leadership is not optimal for an organization that has the desire for collaboration, worker engagement in the organization, and inclusion of workers. Technology development is limited or nonexistent as thinking outside of the box is not often encouraged.
Bureaucratic (By the Book Leader)	This style of leadership focuses on procedure adherence and what has been done in the past, "This is how we have always done it." Work is accomplished and problems are solved through layers of control with the leader; gleaning power from controlling the information and the rate of information distributed.	This style of leadership is not good for organizations that are seeking to foster an inclusive work environment where the talent and voice of others is helpful in developing new processes and products. The inflexibility of leadership in allowing employees to think outside of the box hampers creativity, employee desire to collaborate, and increases fear of mistake and retribution.
Coaching	The coaching leadership style is relatively new in recognition. The coaching leader goal is to get the best out of employees. They devote time to develop and improve the performance and competences of employees.	This style of leadership is not optimal in organizations where employees are not engaged or motivated.
Collaborative	Collaborative leaders support the team approach to problem-solving, are always trying to influence and motivate people to create the best possible outcome, and encourage people to work independently because they value creativity.	This style of leadership can be costly and reduce the independent thinking required of designers and engineers to exercise their innovative skills that are needed for new product development
Charismatic	Charismatic leaders have the innate ability to communicate and connect with others in ways that reach them on an emotional level, to inspire and motivate.	This leadership style has the ability to inspire followers to buy into and assist with goal achievement. Good for innovation; however, the feeling of a lack of leadership may be felt by some professionals.
Democratic	A democratic leader shares decision-making with a person, group, or team. Many refer to this type of leader as participative leadership that encourages participation and the engagement of others. A democratic leader facilitates conversation, encourages people to collaborate and share ideas, and evaluates all information available to devise the best possible decision based on the input.	This leadership style is great for facilitation of teamwork, employee engagement, and building trust in teams and organizations. Supports innovative thoughts and exploration of ideas.

Laissez-Faire	A laissez-faire leader typically defines the goals, policies, and parameters for operations and gives complete freedom to subordinates to make decisions and act on their own to accomplish work. This type or leadership style assumes a hands-off approach when it comes to day-to-day activities, hoping to inspire employees to become leaders.	This leadership style allows employees to use incivility and creativity. This can be useful for new product development, investigation, and development of new technologies.
Servant	A servant leader puts the need of employees, the organization, and customers first, and willingly shares power and extends credit to others for organizational accomplishments. Leadership characteristics include empathy, keen foresight, good listening, and persuasions skills.	This type of leadership is good in environments where employees are expected to imagine, design, and develop new technologies.
Strategic	Strategic leaders have the ability to develop and express a strategic vision for the organization and influence others to support and pursue the vision. This followership influence extends to the way employees are managed. These leaders have the ability to influence organizational members to make decisions that will enhance the short-term and long-term success of a tera or organization.	This leadership style is helpful in technology and product development teams. The strategic nature of these leaders is an asset in guiding the development and implementing of new products and refinement of existing products and services.
Transactional	A transactional leader is one who values order and structure. These leaders struggle in environments where creativity and innovation are expected.	This leadership style is not optimal for research and development organizations or where critical thinking and technology development is of importance.
Transformational	Transformational leaders have the ability to inspire positive changes in their followers. These leaders are passionate, energetic, involved in organizational activities, and focused on the success of every team member.	This leadership style can be used to help teams identify issues and motivate them to find and implement solutions to problems encountered during task performance.
Visionary	A visionary leader sees how things should be and takes the necessary steps to get there. They have a riveting vision for the team, project, or business and can see beyond the obscurity, challenges, and are able to ignore naysayers' comments.	This leadership style is supportive of engineers as they are in the idea stage of product development and serves as a sounding board in flushing out ideas.

TABLE 8.3

Leadership Styles Comparison

Situational Leadership Style	Attributes	Leadership Style Comparison
Telling	Leaders tell their subordinates what to do and how to complete activities and closely supervise performance.	Autocratic Bureaucratic Transactional Laissez-Faire
Selling	Directions and information provided by the leader with extensive communication and collaboration with team members to gain acceptance. Decisions are explained and questions are clarified.	Charismatic Transformational Strategic
Participating	The leader and team members work closely together, share ideas, and facilitate in decision-making, and collaborate closely in problem-solving.	Coaching Servant Collaborative Visionary
Delegating	Many of the decision-making and implementation responsibilities within an organization are passed on to others (subordinates) and take on the role of monitoring progress.	Democratic

Source: (Alston, 2014; Hersey, 2004)

leadership styles listed in the table to accomplish work and lead their teams and organizations. For new product development activities conceivably, situational leadership may be the optimal leadership strategy for leading the project team organization as it encompasses all other leadership styles and offers flexibility in the leadership decision-making process.

8.5   ORGANIZATIONAL CULTURE THAT UNDERPINS AND PROPELS INNOVATION

As discussed in Chapter 7, organizational culture continues to be viewed as an important asset for a company as long as the culture serves as the foundation by which the organization functions. **The functionality of organizational culture includes the way people interact with each other, the way leaders interact with subordinates, the policies that are developed and implemented, the communication channels and their effectiveness, worker engagement, and trust between workers and leadership.** Organizational culture is important to leaders because it provides the context by which an organization and its members function and perform their assigned tasks. According to Cameron and Quinn (1999), organizational culture has a distinct effect on the performance and the long-term effectiveness of an organization. Edgar Schein has been cited as one of the most significant contributors to culture research and his work serves as a cornerstone for others who continue evolution of the theory and importance of culture in organizations. Schein (1992), defines culture as a

> a pattern of shared basic assumptions that the group learned as it solved its problems of external adaptation and internal integration that has worked well enough to be considered valid and, therefore, to be taught to new members as the correct way to perceive, think, and feel in relation to those problems.

Organizational culture studies and literature suggests that a culture can be created to exhibit the creator's values, priorities, and vision and view culture as a key to commitment, productivity, and profitability. Therefore, culture can have a distinct connection to beliefs about organizations such as performance, fairness of leadership, trust between members, the location of authority and flow of information, decision-making practices and processes, leadership style, compliance, fairness and comprehensiveness of evaluations, and employee engagement and motivation.

Activities that can be undertaken to develop and maintain a positive organizational culture which serves as the foundation to support a company competing in a fast-changing technological environment include

- Transparency—Transparency in decision-making and actions taken is necessary to build trust among organizational members, provide the inspiration for members to support the vision set forth by leaders, and improves engagement that facilitates creativity and increased productivity.
- Employee empowerment and engagement—When an employee believes they are empowered to think and utilize creativity in accomplishing their task, they become engaged in supporting the vision and the mission and creative problem-solving to ensure success and increase productivity.
- Communication—Open, honest, and frequent communication is a great tool to use in minimizing the rumor mill, providing workers with accurate and current information, and increase trust among workers and trust in the leadership team.
- Recognize contributions—The act of celebrating success, contributions of others in helping the organization to accomplish goals and objectives provides a sense of gratitude and a feeling of appreciation and being valued.
- Strong worker relationships with leadership and each other—Building relationships is key in forming a bond between people, developing, and retaining trust, and building a sense of team comradery among members.
- Solicit feedback from organizational members—Soliciting and acting on feedback from others creates a collaborative environment when employees feel that their input is valued and necessary to the success of the organization.
- Facilitate a team atmosphere—Developing a team atmosphere unites employees and channels their focus on a common goal that will serve the business in a positive way. A team atmosphere inspires a sense of belonging, increased collaboration, creative thinking, and builds trust among members.

- Embrace an inspire creativity—Creativity encourages new ideas among workers and is necessary to develop new process and products and discover novel ways of solving problems.
- Embrace diversity—Sets the stage for improved employee performance because individuals from diverse backgrounds can offer different skills, talents, and experiences that a company can leverage.
- Recognize that humans are fallible—Because humans are fallible, people will make mistakes from time to time. It is important when these mistakes are made, they are used as "teachable moments" and employees are treated in a respectful manner. When evaluating the mistake, processes and systems involved are evaluated and modified as needed to prevent the mistake from reoccurring.
- Build an inclusive environment—This type of environment provides the assurance to workers that they are valued, including their differences, and their differences are viewed as an asset to the company.
- Leaders set the example by walking the talk—Leaders must set the example by following the policies and procedures set by the organization and respond to organizational members in a manner that they expect members to respond. Walking the talk means doing what is expected of others.
- Create a shared vision—A shared vision is necessary for a company to achieve its mission. A shared vision can serve as an energizer for a team or organization and facilitates a collaborative environment
- Respect and trust—When people are treated with respect, they are more likely to extend respect and open themselves to trusting others. Trust will develop as a byproduct that can create a bonding relationship that will support team cohesion and impact improved performance

A company's culture can be a viable source of sustainable competitive advantage, guide behaviors, facilitate team cohesion, support diversity of opinions, promote inclusion of workers from differing backgrounds, and improve overall organizational health and performance.

8.6   THE LEADERSHIP IMPERATIVE: LEADERSHIP ADAPTATION AND CHARACTERISTICS

Leadership is important in driving the success of organizations, teams, and people. Key leadership imperatives include the following:

- Developing and charting the vision for the organization
- Facilitating a culture of trust that embraces diversity and inclusion
- Driving improved performance of people, teams, and the organization
- Coaching, mentoring, and developing people
- Competency in managing the work and leading people
- Leading with honesty, trustworthiness, and fairness
- Inspiring loyalty and trust and influencing others to follow

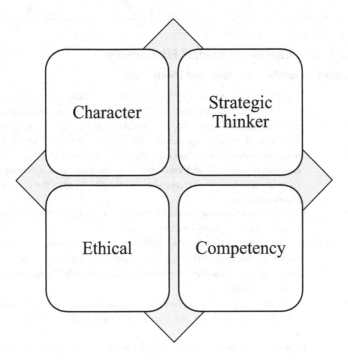

FIGURE 8.1 Effective Leadership Characteristics.

Leaders who are able to effectively function and embody the imperatives are generally strategic thinkers, have admirable character, conduct work and all interactions in an ethical manner, and are competent as a leader. These effective leadership characteristics are shown in Figure 8.1 and further discussed in Sections 8.7.1–8.7.2.

8.6.1 LEADERSHIP CHARACTER

Leadership charts the course of an organization and is responsible for success or failure in terms of productivity and viability of the team. As such, it takes a leader possessing the right embedded characteristics, which are visible and recognized by everyone they come in contact with to yield organizational success. There are many facets of thoughts on leadership and what it takes to demonstrate successful leadership. **The character of a leader is the aspect of a person that will allow others to engage in a trusting relationship that will allow business processes to transpire and grow.** A person's character encompasses their mental and moral attributes that are visible through their actions and interactions and are key in relationship building and maintenance. Important elements of an individual's character that can help a leader achieve success in the workplace are listed in Figure 8.1. There are other attributes or elements of a leadership character that are used by effective leaders; however, the elements are sub-elements of the primary elements shown in Table 8.4.

TABLE 8.4

Character Role in Charting Individual Effectiveness

Character Element	Contribution to Individual Effectiveness
Respect	Treating others with respect will yield respect for the extender.
	Workplace Environment Benefits (WEB): when leaders are respected by workers and each other, comrade in the work environment increases; allowing tolerance for and the ability to accept human differences and includes those differences in decisions that are made and cultivation of the culture of the organization; enhances the ability to deal with conflicts and disagreements, achieving a win for everyone where feasible and when not feasible the party that conceded still supports the team.
Responsibility	Do what you are supposed to do, plan in advance, exercise self-control, and demonstrate accountably for your actions
	Workplace Environment Benefits: facilitate the desire and practice of making good choices for members of the organization and the organization; demonstrate self-control and accountability for actions, and spoken words facilitate trust among organization members
	Think before you act • Be accountable for your words, actions, and attitudes • Set a good example for others to follow in taking responsibility for their words and actions
Caring	Demonstrate kindness, empathy, fair heartedness, compassion, and gratitude for others.
	Workplace Environment Benefits: a caring leader that shows that they are concerned with the safety and well-being of workers is able to tap into and facilitate trust, enhances desire by workers to go the extra mile to ensure team success, and reduces or eliminates the fear of admitting mistakes when they occur.
Stewardship	Doing what is right to make your home, community, and the world a better place for the present and into the future.
	Workplace Environment Benefits: responsibility and care are given to the type of work environment created for workers and care in the eliminating or reducing potential negative impact to the environment (upholding all work laws).
Core Values	Underpins individuals' principles and standards that guide behavior.
	Workplace Environment Benefits: core values serve as the guide to behaviors. Leaders who demonstrate the appropriate behaviors (good judgement, humility, high integrity, transparency) are successful gaining willing followers who are supportive of the goals of the organization.

The character of a leader is also a key attribute in culture development and sustainment, facilitating diversity and inclusion among members, and trust development and sustainment in organizations. Consider the following scenario of how the character of a leader can be applied to problem-solving in organizations. Company A produces a special kind of widget, Widget A, that is a critical component of the steering system for the automobile industry. The widget is similar in shape and size as another widget, Widget B, that is used in assembling bicycles. A large order was placed by one of their top customers, Company X, for the Widget A to be installed in all their automobile steering systems. The worker filling the order inadvertently packaged and shipped the Widget B. Two days later while validating Widget A, a quality control inspector identified the discrepancy and quickly realized that the wrong widget was shipped to Company X. The employee immediately notified management of the mistake and communicated how distraught they were for making such an error. Management quickly assembled an investigation team to evaluate why the mistake was made and what needed to be done to avoid reoccurrence. Employees performing the task of filling orders and maintaining inventory were actively engaged in and supported the investigation.

Once notified, the management contacted Company X to make them aware of the wrong shipment and apologized for the mistake, informed leadership that the correct part was being expedited for shipment and should be received by the company within two business days. In addition, Company A will pay to have the parts shipped back to them for recycling or reuse to avoid disposal of the parts and help to defray some of their loss through refilling the order. Company A further informed the company what process changes were made to avoid future mistakes. They further offered to compensate the company for any rework or cost incurred as a result of incorrect shipment. Company A also assembled an evaluation team to review current internal processes to determine if changes were warranted and ensure mistakes of this type are not repeated. Upon being made aware of the process weaknesses, the manager met with the employees and communicated that the previous process was flawed and contributed to the mistake made in filling the order. Table 8.5 lists some actions that a leader may take during the performance of their roles that is directly associated with their character. The table lists some, but not all of the possible benefits yielded from the scenario as a result of the character of the manager that was evident by their action.

It is clear from the analysis of the scenario that **the character of a leader can make a significant difference in the way work is accomplished, the way communication flows, the relationship the company has with workers, internal and external customers, as well as the overall viability of a company.**

8.6.2 ETHICAL LEADERSHIP

How can an ethical leader be recognized in organizations? These leaders always do the right thing even when it is unpopular or inconvenient. Ethical leaders generally display the traits listed in Table 8.6, at a minimum, in their daily interactions and decisions they make. An ethical leader has high values and does not tolerate unethical behaviors; therefore, they expect employees to mimic ethical behavior during their interactions.

Ethical leadership behavior can eventually become contagious and the norms for followers to emulate. In such a case, the number of unethical activities taking place in an organization is reduced or at best becomes extinct. Customers engaging in

TABLE 8.5

Leadership Character in Action—Scenario

Character Element	Workplace Environment Benefits (WEB)
Respect	The action of the leadership for Company A demonstrates a number of characteristics that will lead to respect such as the action to contact customer and take responsibility for shipping incorrect part; workers were included as a part of the investigation team and were not immediately blamed for the mistake.
Responsibility	Leadership took responsibility for the shipment mistake internally and to internal customers and did not immediately blame workers; took responsibility in assembling a team to determine cause of the shipment mistake, determinate mitigation action, and ensure implementation of preventive measures
Caring	Demonstrated care and concern for employees by letting them know about the flaw in process once discovered; trained workers on the new process to ensure consistency in implementation
Stewardship	Having the unneeded products returned to the company for redistribution to avoid products from being disposed of as waste
Core Values	Good judgement was exercise based upon the actions taken to preserve and strengthen customer relationships and internal relationships with workers. Values such as humility, honesty, integrity, transparency, etc.)

business with these organizations are less concerned about being taken for granted and expect that the products or services provided are of high value that will meet and most times exceed their expectations.

8.6.3 COMPETENT LEADERSHIP

How would you define competent leadership? Many would include in the definition technical knowledge and skills in the subject the leader is responsible for leading. A leader can be technically competent and not possess the skills of a competent leader. There are several key characteristics that competent leaders display in their daily roles and interactions. Key characteristics of a competent leader are listed in Table 8.7.

Competent leaders have the ability to influence others to be the best that they can be and follow leadership, impact organizational culture, and they always exercise the ability to do what is needed. Leaders who display the key characteristics listed in Figure 8.5 are equipped to successfully lead their organizations and create a work environment that will inspire others to improve their abilities and their contributions to the organization in meetings goals and objectives.

8.6.4 STRATEGIC THINKING LEADERSHIP

Have you ever witnessed an organization where members are unsure of goals, priorities, which project is important, customers' needs, etc.? If so, you have witnessed an organization without a strategy and without a strategic leadership team. As stated by Sun Tzu, "Strategy without tactics is the slowest route to victory. Tactics without strategy is noise before defeat."

TABLE 8.6
Leadership Traits of Ethical Leaders

Common Ethical Leadership Traits	Description in Action
Honesty	It is understood that an ethical leader will also be honest, transparent, and trustworthy. Honesty is of particular importance for leaders with the desire to be viewed as ethical and effective in performing their roles as leaders. Followers will willingly follow leaders they trust.
Respect for All	Ethical leaders are attentive listeners, value the contributions of all members of the organization, value diversity, and facilitate inclusiveness of workers. Treating people with respect is cyclical, meaning respect will be given to the extender.
Leadership by Example	Ethical leadership is about walking the talk. The expectations that an ethical leader has of themselves and employees is to conduct business while leading by example.
Just Treatment	An ethical leader always demonstrates fairness in decision and the way people are treated. Avoid having favorite employees and embrace diversity and include everyone equally. Employees are not fearful of being treated unfairly based on such factors as handicap, gender, race, nationality, etc.
Intolerant of Unethical Behaviors	An ethical leader expects employees to follow their lead and do the right thing with every encounter even when it is not easy or palatable to others.

TABLE 8.7
Competent Leadership Characteristics

Leadership Characteristic	Benefit to Organization
Trustworthy	Some attributes that trusted leaders bring to an organization include empowering workers to think and make decisions, facilitating productivity because they are inspired to create and produce without fear of failure, can play a part in the decision of an employee to remain with the organization or to seek employment elsewhere (Alston, 2014).
Ethical	Ethical leaders set the example for others in the organization to conduct business in an ethical manner which will have a positive impact on members and customers of the organization. Ethical leaders are open and honest and will always honor their words. Employees and customers respond well to and trust ethical leaders.
Courageous	Courageous leaders are not afraid to make decisions even in the face of controversy. These leaders have high expectations and are not afraid of asking large request of others and are willing to explore uncharted and unproven territory. This attribute is key in organizations where change is needed, such as reinvention, product changes, or organizational realignment.

(Continued)

TABLE 8.7
Continued

Leadership Characteristic	Benefit to Organization
Inspiring	Inspiring leaders are positive thinkers and speakers who can always see the bright side of a situation and as such are able to impact the views and the aspirations of their team.
Humility	Leaders who lead with humility will quickly admit mistakes, give credit to others, are respectful to everyone, and accepting of feedback from others. These actions are instrumental in gaining respect and trust of workers and customers. In addition, employees are not fearful of admitting mistakes when they occur.
Inclusivity	Employees working in a culture and for a leader that embraces the practice of inclusivity, tend to practice inclusivity themselves. These leaders have policies in place to facilitate equal access to opportunities and resources for everyone, and they are more apt to be engaged in the business to ensure success. In addition, they are actively involved in problem-solving and contributing to seeking ways of doing more efficiently.

Strategic leadership just does not happen; it is planned and nurtured. Deploying a strategic leadership team begins with the leadership selection process. When selecting leaders, there must be a process used to identify and put in place the leaders that will complement the strategy and execution of the organization. Some questions for consideration when identifying and hiring a strategic leader include

- Describe some of the important values you possess as a leader
- How do you influence others to accept your ideals and direction and follow?
- How much time per month do you dedicate to strategic planning?
- How would others who you have led describe your leadership skills as it pertains to developing and implementing a strategy?
- How would you resolve conflict among team members in the event of a disagreement?
- How to you lead through change management?
- What method do you use to set long-term and short-term goals for your organization, and how often to you verify status?
- What factors do you consider when building an action plan?
- How do you measure the effectiveness of a strategy?
- Describe an occasion when you proactively identified and resolved an issue in the work environment.
- What activities or steps have you engaged in during your career to become a strategic thinker?

The interview questions should be posed in a such a way so that an evaluation of whether the candidate embodies the characteristics that will move the organization forward in a positive way is revealed. The list of questions above are examples of

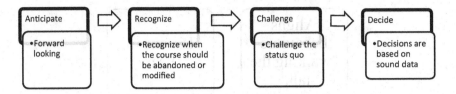

FIGURE 8.2 Leadership Mental Model for Progress.

some of the questions that may be included in the interview process. Other screening activities may include asking the candidate to comment on a scenario, giving them real problem to solve, or asking them to comment on a flawed strategic plan. A strategic leader must be able to anticipate changes and issues, recognize early when the strategy or path to success needs to be modified or abandoned, challenge the way things have always been done, and make decisions based on sound data that will move the company forward and bring along organizational members (followers) and customers. The thought pattern of a strategic leader is orderly and implemented in an orderly pattern much like the road map that is presented in Figure 8.2.

Strategic thinking leaders do not conduct business, formulate relationships, or make decisions while "flying by the seat of their pants" or while in reactive mode. They take the time to ensure that their decisions will yield the best possible outcome. Some of the considerations that a strategic leader will include in their decision-making process are listed below:

- Takes the time to understand the issue at hand
- Collaborates with others
- Listens attentively and respects the opinion of others
- Evaluates the pros and cons of their decision
- Evaluates the potential impact of their decision on others (employees, customers, stakeholders, etc.) and the environment
- Considers the impact of the decision on current and future company activities and initiatives

Although a leader may seek data from others in order to make a decision, once the decision is made, that leader owns the decision. In the event the decision was not fruitful for the situation or the organization, a true leader will not blame others. They will take responsibility and make whatever changes are warranted to render success. In the event the leader proceeds to blame others, they can expect to lose respect and trust of colleagues, employees, and potentially customers.

8.7 COMPETENT LEADERSHIP SELECTION

Competent leadership is not always easy to find. It has been my experience that there are times when leaders are chosen because they are good individual contributors, and they are well liked by employees, leadership, and the customers, etc. These are acceptable reasons to consider when identifying competent leaders; however, they are not the only factors that must be explored. Additional items that must be explored

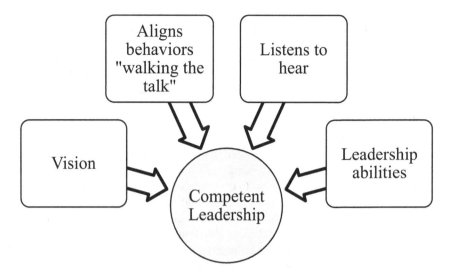

FIGURE 8.3 Leadership Competency.

that are of significant importance are listed in Figure 8.3. and illustrate the importance of these attributes to a leader being viewed as competent by their followers, customers, and stakeholders.

It is important to know that leadership competency has little or nothing to do with their technical knowledge in the area they are responsible for effectively leading, although oftentimes the mistake is made to connect competency with technical knowledge. Alston (2014) describes leadership competence in terms of influence, ability, impact, expertness, knowledge, and the ability to deliver what is needed to advance the organization. Four key attributes that a competent leader must embody and display are discussed in more detail below:

- Vision—A vision is important in focusing members of an organization on what is important to the company. A leader without a vision may be incapable of inspiring followers to follow.
- Aligned behaviors walking the talk –Followers are more likely to emulate the behavior of their leaders; therefore, leaders who exhibit the behaviors that they are expecting of members of the organization are more likely to see those behaviors manifested.
- Listens to hear—Effective listening skills provides the gate to information assimilation and problem-solving in order to hear what communication is necessary for the leader to strive for during every interaction to fully understand what others are wanting to communicate. Leaders with good listening skills are perceived to be caring, confident, engaged, and mindful of others because of the attention provided during attentive listening.
- Leadership abilities—In order to be viewed as an effective leader, an individual must have the necessary skills and abilities to function effectively in the role as viewed by others. Leadership skills and abilities are different

from technical skills. An individual can possess appropriate technical skills and lack the leadership skills needed to lead a team or an organization.

8.8 APPLIED LEARNING

1. Name and discuss the characteristics of competent leadership. Discuss how each characteristic can be successfully utilized by organizations.
2. List and discuss the four-leadership competency. What role do these competencies play in the support of inclusive product design?
3. List and discuss at least five questions for consideration when identifying and hiring a strategic leader. Discuss the importance of the selected questions in identifying and selecting the appropriate leader for a design team.
4. What steps can an engineering manager take to foster and nurture designers within their technical disciplines?
5. Discuss the meaning and the importance of the following statement on new product design; "The character of a leader is the aspect of a person that will allow others to engage in a trusting relationship that will allow business processes to transpire and grow."
6. Discuss the role organizational culture plays on team dynamics and development of new products that will be used by a diverse population.

BIBLIOGRAPHY

Albrecht Simon L., Perceptions of Integrity, Competence and Trust in Senior Management as Determinants of Cynicisms Toward Change, *Public Administration and Management an Interactive Journal*, Vol. 7, no. 4, pp. 320–343, 2002.

Alston Frances, *Culture and Trust in Technology-Driven Organizations*, CRC Press Taylor & Francis Group, 2014.

Cameron Kim S. and Quinn Robert E., *Diagnosing and Changing Organizational Culture*, Addison-Wesley Publishing Company Inc., 1999.

Hersey Paul, *The Situational Leader*, The Center for Leadership Studies, Inc., 1984.

9 Individual Readiness for Engagement

9.1 INTRODUCTION

The activities performed by individuals are the pathways by which organizations achieve their goals and overall business success. A recognition by leadership that each individual employee has a role that is important to the organization achieving the goals and mission is important. This public recognition by leadership provides the glue that bonds and enhances acceptance of and support of organizational goals because employees believe that they are valuable to the organization and take pride in their work. Not only must leaders recognize the importance and contribution of the individual employee, but they must also demonstrate this belief in the way they communicate with and treat workers. We often hear leaders say the greatest asset of the company are the employees. However, if the employees were asked if they believe that they are valued, the answer oftentimes is no because they do not believe that they are valued by leadership; many employees would state management is focused on making production goals and minimizing impacts to themselves. Leadership plays a critical role in rendering individuals ready for engagement in the business of an organization. A significant factor in engagement is individual trust in their leaders. Demonstrating honesty and concerned for the well-being of workers are at the core of employees recognizing their value and trust within organizations.

The first step in preparing an organization for engagement is to determine the current engagement state, establishing a baseline of organizational engagement. This is accomplished by obtaining feedback from organizational members along with reviewing current policies and practices. Feedback can be obtained through administering employee engagement surveys and/or conducting focus group discussions. Engagement surveys are popular, and many companies including Fortune 500 companies, conduct these surveys on some frequency to gauge employee perceptions and to better understand the culture of the work environment. The use of focus group discussions is less popular; however, focus group discussions can also provide valuable information on engagement status. Engagement surveys have gained popularity in the 21st century because they are a low-cost method that has been successfully used in assisting leadership in understanding the perceptions of employees with regards to their workplace conditions that can have a direct impact on their performance.

Preparing an organization for full engagement is not an easy task and can be time-consuming and yes, frustrating. Organizational engagement should be an integral part of the overall business strategy for a company since engagement enhances the ability of the leadership team to gain access and use the creativity and diverse thoughts of all members of an organization. In an organization where engagement is a priority, the work culture has the potential to become a competitive advantage,

DOI: 10.1201/9780367854720-11

retainment of talent is optimized, and performance has the potential to reach peak performance. The term peak performance refers to the point where an organization has implemented streamlined processes that has a positive impact on the business, reduces rework, and human error is anticipated and managed, and employees are actively engaged in the business (Allen, Alston, and Dekerchove, 2019). The peak performance model introduced by Allen, Alston, and Dekerchove is a good model to use to understand and facilitate engagement and improve organizational performance. The model is based on four components that can assist an organization in achieving and maintaining excellence. These components are lean, culture, human performance improvements, and operational excellence.

What does full engagement mean for an organization? How can engagement be achieved?

Full engagement refers to the point at which everyone in the organization has bought into, demonstrates support for the vision, and is actively engaged in ensuring the goals set to achieve that vision are accomplished with quality and precision. Engagement can be achieved when the organization is built on a culture that supports and welcomes diversity and inclusion of all members; attempts are made to embrace and understand the differing ways that various people from different backgrounds approach decision-making and problem-solving, led by a trustworthy leadership team that is strategic, encourages employee involvement, creativity, and demonstrates genuine respect for others.

Today companies spend a significant amount of time strategizing on ways that employee engagement can be increased in their organization because they realize the benefit of having engaged employees. When seeking to increase employee engagement in an organization, the following must be considered at a minimum:

- Employee engagement must begin and be included as a part of the business strategy with a focus on employee recruitment and retention.
- Engagement of workers should have a primary focus of achieving the desired results to help the organization achieve peak performance. Therefore, accountability must be a part of the equation for workers and leaders.
- A mechanism to provide a means to measure business outcomes and communicate performance throughout the company. Engagement tends to increase when workers are able to connect the tasks that they perform with the overall performance of the company.

According to Alston 2017, there are five major benefits of employee engagement that include improved morale, increased productivity, elevated level of trust between leadership and employees, increased team cohesion, and innovative thinking. These benefits are essential for organizations that engage in high technology, product development, and research and development. These benefits along with some of the expected results are shown in Table 9.1.

In order to prepare an organization for "full engagement" attention must be paid to getting every participant within that organization focused and onboard where feasible. Recognizing that full engagement may not be achieved, working toward

TABLE 9.1

Engagement Benefits and Results

Engagement Benefit	Expected Results
Improved Morale	Increased morale is seen in employees that are engaged and are treated as if they are important assets of the company. Additional key benefits of improved morale include increased retention rate and the willingness of employees to go the extra mile to assist the organization in achieving goals.
Increased Productivity	Engaged employees are invested in business outcomes; therefore, they work hard to ensure that the vision of the organization becomes a reality. Productivity increases occur because engaged employees use their creativity to develop new technologies, products, and services to help support business needs.
Increased Trust	Trust is viewed as one of the single elements that can help improve engagement. Trusting people allows themselves to be open to new thoughts, ideals, and ways of doing things. More importantly when an individual trusts, they will allow themselves to be influenced by the individual that they trust.
Team Cohesion	Team cohesion can lead to a cohesive work environment that can improve employee job satisfaction and increase involvement in support for the business.
Innovative Thinking	Innovative thinking can lead to new product and technology development, as well as different and more productive ways of conducting business.

gaining full engagement can be beneficial in itself in assisting the company to gain the support and corporation of workers in goal achievement. Leadership must serve as the example for what is expected from others, this means that leaders are modelling expectations for performance. Also, peer pressure in this case, may move other workers to become more engaged especially when they can see that the voice of workers that are engaged in the day-to-day business is embraced and considered during decision-making by leaders. Employee engagement, or the lack thereof, can have a significant impact on the success or failure of organizations, the services they provide, and the products they develop and produce.

9.2 THINKING STYLE INFLUENCE ON READINESS

Thinking is at the foundation of everything we do and is the process that every human being uses to approach and find solutions to problems, make decisions, contemplate, develop new ideas, and unleash their creativity (Kallet, 2014). The way and the process of thinking determines how an individual approaches dealing with everyday issues and interacting with others. In conducting the act of thinking, various thinking styles are used by individuals. Thinking style is being defined broadly as the mental structures that permit individuals to process information and solve problems in a defined way (Belousova and Pishchic, 2015). To compete and advance in the global marketplace, corporations are seeking to hire, develop, and retain employees who are critical and strategic thinkers. Therefore, training institutions are focusing more on developing and providing training to workers to improve thinking skills.

9.2.1 CRITICAL THINKING

According to Kallet 2014, everyone can become a critical thinker recognizing that some people have more of an inclination toward thinking critically than others; everyone has the capability to improve their thinking when approaching problems; and critical thinking skills can be taught and learned. Critical thinking is the analysis of a series of facts to inform a decision. Critical thinkers have an innate ability to understand how ideas are connected and evaluated; understanding whether the information presented is relevant in solving issues can identify gaps in data or errors in.reasoning that leads to cohesive defendable conclusions. Some characteristics of critical thinkers include

- Behaves in a well-informed manner and has the desire to continuously learn
- Demonstrates the ability to readily identify situations where focused thinking is warranted
- Demonstrates a strong belief in their analytical and reasoning skills
- Are curious of their environment and issues as they evolve
- Embraces a diversity of viewpoint, beliefs, and opinions of others
- Recognizes any biases that they may have that can impact their judgement
- Demonstrates objective and fair interactions
- Formulates judgement or opinion only after all of the facts are known and analyzed
- Demonstrates the ability to re-evaluate or abandon their belief when presented with data that does not support their conclusions

Critical thinking skills do not generally come naturally; however, the skills can be taught or modelled as seen used by others (Rutherford Albert, 2018). Courses designed to teach critical thinking skills are available for managers to take advantage of as a tool to train staff to improve individual employees and enhance organizational performance. This is good news for leaders because companies need people who are capable of thinking critically and can help identify, analyze, and develop novel solutions to problems. Figure 9.1 shows some key aspects that encompass critical thinking for most individuals that should be mastered. Each aspect has a significant role in the process and is further defined in Table 9.2.

Solutions to problems, development of new technology, products, and services continues to be a significant role that is performed by workers. In order to successfully implement the practices of developing inclusive and speculative design, and to draw conclusions that are thoughtful and effective utilizing the aspects listed in Figure 9.1, critical thinkers make use of asking logical questions and targeting key factual details that provide substance to the process of thinking and yields logical and defensible decisions. The Critical-ACT approach reduces the prospect of drawing or concluding a point of view that is inaccurate, illogical, unsupportive, and non-defensible. If one were to imagine a world without critical thinkers, technological advances would be stifled, minimized, or non-existent. Workers and professionals who have mastered the aspects of thinking are generally considered key to employers and the success of their companies' progression, sustainability, and existence.

There are several mental actions that are completed during critical thinking. These actions may be differently performed depending on the skills and style of

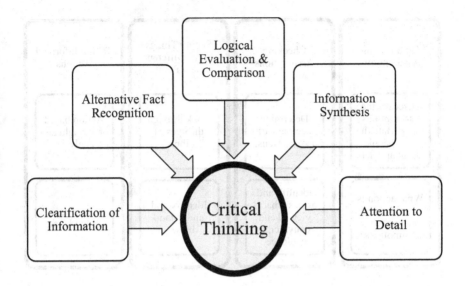

FIGURE 9.1 Key Aspects of Critical Thinking (Critical-ACT).

TABLE 9.2
Aspects of Critical Thinking Defined

Critical Thinking Aspect	Role in Critical Thinking
Attention to Detail	Attention to detail is important to critical thinkers because it means that the individual has the capability to notice, retain, and keep track of details, follow instructions, and carefully evaluate complex problems with a high level of precision.
Logical Evaluation and Comparison	The ability to logically evaluate analyze and compare process flow, data, and scientific facts enhance critical thinkers' ability to reach defensible responses.
Alternative Fact Recognition	Alternative facts move the thinking process from reality into a realm of thinking that can and will distort the outcome of a decision that led to defensible actions. The capability to recognize what is truthful and accurate provides a strong basis for data synthesis and decision-making.
Information Synthesis	A keen ability to synthesize information enhances the ability to combine ideas and develop a clear understanding of information and composite data.
Clarification of Information	The ability to clarify information allows one to make sense of challenging and complex issues.

different individuals performing the actions. Here is where it is important to remember that people are different and different does not mean they are wrong or undesirable. Diversity of thought is a great way of introducing various views and avenues to solve problems and interject new ideas and ideals that can lead to the development of solutions, services, and technologies that are accepted and compatible for a larger

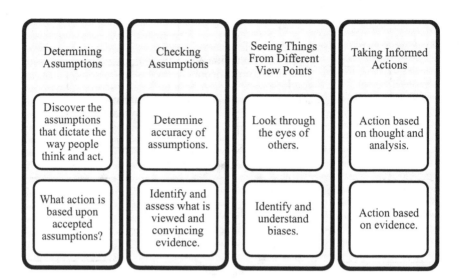

FIGURE 9.2 Critical Thinking Paradigm.

and more diverse global population. According to Brookfield (2012), critical thinking happens when the actions listed in Figure 9.2 are complete. These actions together form the road map that leads to, and completes, the critical thinking process. A key point to remember is that critical thinking helps an individual to look at situations differently (completely and thoroughly) and the process or behavior is not automatic; rather it is deliberate. This means that to engage in the process of critical thinking, a deliberate choice must be made, and the accompanying behaviors follow.

9.2.2 STRATEGIC THINKING

Strategic thinking is the second key thinking process that is important in performing work where technology advancement, product development, and research and development takes place. This style of thinking can provide a competitive advantage for companies as they prepare for present and future viability. **Effective strategic thinkers are believed to have mastered skills that are important to develop and implement strategies and have the foresight needed to change direction and modify strategies when appropriate.** These skills include the ability to think critically and to anticipate and look beyond what is apparent. A skilled strategic thinker has the ability to develop effective plans that will propel the organizational vision into reality, having the sustainment mechanisms in place necessary to withstand the test of time. In preparation for developing and the implementation of an effective strategy, the strategic thinker makes use of the aspects of strategic thinking that include critical thinking that are listed in Figure 9.3 and discussed in Table 9.1.

These aspects of strategic thinking are important for paving the way for informed decision-making that can withstand challenges, discovery of new products and services, entering new markets, hiring, and retaining talented resources. The role each aspect plays in the strategic thinking process is discussed in brief in Table 9.3. The

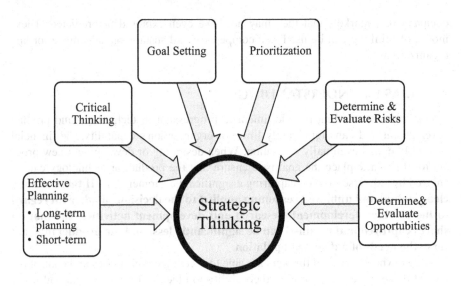

FIGURE 9.3 Aspects of Strategic Thinking.

TABLE 9.3
Aspects of Strategic Thinking Defined

Strategic Thinking Aspect	Role in Strategic Thinking
Goal Setting	Great tool to provide focus on where the organization is heading
Effective Planning—Long- and Short-Term	Effective planning is key to achieving goals utilizing information and forecasting and formulating reasonable prediction for current and future potential activities.
Prioritization	Helps participants focus on what is important to the organization to achieve the goal
Determine Risk	Identifying and knowing the exposure risks allow for informed decisions and aids in risk avoidance, mitigation, and acceptance
Determine Opportunities	Knowing and focusing on known opportunities provides the details needed to determine the path the company will take in terms of business ventures and markets entries
Critical Thinking	The problem-solving skills provides the foundation for decisions and the plans that supports strategies.

elimination of any of the listed aspects is sure to have an impact on the quality of the decision and strategy employed. In the business environment the impact can range from development of flawed technology and products to a loss of resources (financial and human) or even the loss of company credibility. The impacts can also place a company into a nonrecoverable posture, resulting in bankruptcy and business closure.

When a company is staffed with employees and leaders who are skilled in critical and strategic thinking, they are able to imagine, develop, produce, and propel the company into markets that they may not have even explored or predicted. These modes of thinking can be used as a competitive and sustainable advantage for any organization.

9.3   BIAS IN UNDERSTANDING

A bias in understanding is a detriment to progression in technology and product development and can significantly limit an organization's capability, the financial bottom line, and eventually reputation. When designing or developing a new product for the marketplace the goal is to ensure that the product or technology can be enjoyed by a diverse market, capturing a significant customer base. **If there is not a clear factual and unbiased reasoning applied to the decisions made with regard to the product development, research and development activities, the market shares captured and retained can be significantly less, and the product may not serve the needs of a diverse population.**

People who are critical thinkers are able to recognize their biases and control the impact these biases may have on their ability to objectively review data and make decisions. Critical thinking requires the thinker be open to the topic or present data without injecting preconceived notions and without biases. The work environment is made up of people from various backgrounds, knowledge, experiences, and perceptions. These workers bring to the workplace any biases they have learned and can communicate and influence workers to these biases through their daily interactions. Depending on the biases and their impact on workers, a variety of responses or actions can be observed from workers, such as not being willing to accept thoughts or feedback from others and dismissing any discussions that do not support their biases. Individual biases can have significant impact on an organization and its culture. Some of the potential impact include

- Stifling creativity. Workers are less engaged in problem-solving and creative thinking.
- Limiting employee engagement in utilizing their skills and knowledge
- Recruitment and retention of qualified workforce
- Difficulty in staffing cohesive and diverse teams

The potential impact of bias in understanding on technology advancement and product development can include

- Product development that exhibits flawed performance
- Reduced market shares that the company is able to secure because the product is accepted only by a small segment of the population
- Impacts to diversity and inclusion of ideas being offered, accepted, and utilized

Biases have a direct impact on the way we think, solve problems, interact with others, and the list goes on. When working in the area of new product development and design of new technology, these biases can impact the end product and limit the potential success of a product. You may question how this limitation can occur. Let us consider the following scenario and the impact of biases.

9.3.1 SCENARIO

Three male design engineers from the same region of the country are developing a product that will track the pulse of individuals with certain health conditions. The pulse monitoring device has two critical areas of connection: one on the individual telephone and the other on the body. The sensor connection requires a distance of no more than two inches apart to obtain accurate results. This design parameter was selected because the engineers had an understanding the 99% of the population carried their cell phone in their pockets. The design was finalized and sent for mass production. After about six months of market time and review of the sales data, it was recognized that 75% of the product was purchased by males and 25% purchased by females. The company was unable to ascertain if the products purchased by females were given to the males as gifts. Here are some of the possible impacts of biased assumptions and understanding that were introduced at design conception and continued through product development.

- Lack of a diversified design team
  - Men have different biases and beliefs than females.
  - Team members from the same area or location may have ingrained biases that are representative for people of the same location.
- The assumption that people carried the cell phones in their pockets
  - Many females tend to place their cell phones in their purses or directly carry the cell phone.
  - Many females may not wear pants or pants with pockets on a daily basis.

Having a bias in understanding just does not magically appear. Developing a bias takes time and exposure to information and behaviors that are in support of nurturing the bias. There is a progression that led to one having a biased understanding. The progression from unconscious bias to bias in understanding of individuals is shown in Figure 9.4.

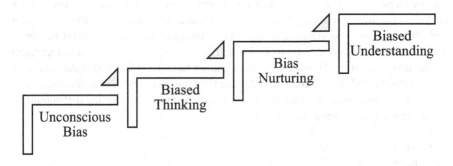

FIGURE 9.4 Path to a Biased Understanding.

These biases are not passed down to an individual at birth; they are learned stereotypes that become automatic and mostly unintentionally used. That process begins with the individual having unconscious biases that may have been formed early in their lives from being exposed to various stereotypes. These stereotypes are nurtured and become deeply ingrained and directly impact decisions and behavior of an individual.

9.4 FOSTERING AND NURTURING THE CRITICAL AND REFLECTIVE PRACTITIONER

The role of a leader is expansive including organizational design, setting policies, and ensuring procedures are developed to outline expectations for the way the business should be conducted, providing training, and mentoring staff. In addition, the leadership team has the role of modelling and facilitating the desired behavior they would like to see in organization members along with designing and facilitating the appropriate culture. The strategy used to develop and nurture critical and reflective practitioners must include providing targeted and appropriate training, coaching, and mentoring.

The development and retention of practitioners that are key to technology development, engineering design, research and development, and other roles that are necessary to provide products and services to a global economy is challenging especially in today's climate with and above average economy having a plethora of technical opportunities available for skilled candidates. A reflective practitioner is being viewed as one who has a continual desire for learning, at some periodicity takes the time to reflect on past work activity and performance and determines areas of improvement and can critically evaluate themself and engage in learning activities that will close the gap in knowledge to improve performance.

9.4.1 TRAINING

Training must be a part of the strategy to ensure practitioners are equipped to manage the jobs they are asked to perform. Training should provide practical experience and practicality that complements the technical knowledge gained from attending learning institutions. Recognizing that early career technical professionals are not trained on how and when soft skills are appropriate for use to ensure success of their business encounters, leaders should ensure that the appropriate soft skills courses are available, and professionals are trained. Organizations leadership should be attentive to ensuring that technical professionals receive the skills needed to render them technically competent. However, the same cannot be said about the soft skills needed for success. Soft skills are recognized more today as being critical to the success of professionals in a technology-based environment. Some of the training that should be considered for all professionals include

- Effective communication
- Implicit bias
- Diversity and inclusion

- Managing difficult people and conflict resolution
- Leadership
- Effective listening

The list above represents some of the skills that have been noted to be lacking among many engineers, designers, and technical workers to cultivate and nurture workers so that they can reach their highest potential. Other training that are considered key are critical thinking and strategic thinking skills.

9.4.2 MENTORING AND COACHING

Many Fortunes 500 companies use mentoring and coaching as an integral part of their employee development and retention strategy. These processes have been successfully used by many organizations to build and strengthen the skills of professionals and are viewed as low-cost measures that can yield maximum benefits when implemented properly. Mentoring is an activity that makes use of pairing less experienced and knowledgeable professionals with more knowledgeable and experienced professionals. A mentoring relationship is rewarding for both parties from a personal and professional standpoint.

Mentoring provides an opportunity for participants to develop communication and professional skills, advance career, and discover different approaches to solving problems and approaching issues as they arise. The mentoring process focuses on the future and therefore is a tool that leaders mostly used to prepare critical staffing for future assignments and roles. Some potential benefits of having an effective and active mentoring program include enhanced employee engagement, increased job satisfaction among employees, increased employee retention, and enhanced knowledge and skills.

Coaching is a personal development activity where an experience and knowledgeable individual supports another who is less experienced or knowledgeable in the discipline or skills, providing guidance and training opportunities. The process of coaching generally focuses on enhancing knowledge and skills for the present. The use of coaching has a negative stigma for some as it has been used to help correct performance issues when professionals fail to meet the expectations of the roles and responsibilities within their organization.

9.5 TOOLS AND TECHNIQUES: BEST PRACTICES

In developing or identifying tools and techniques that can be successful as best practices for preparing professionals for engagement in organizational matters, benchmarking is a good tool to use to identify the strategy, processes, and practices that have been used successfully by others when seeking to improve employee engagement. Benchmarking is a process used to compare an organization's products, services, and processes against those of organizations known as recognized leaders in that area. This process has been and continues to be used by many companies to successfully improve their processes, systems, and products. In addition, through benchmarking, a company can identify areas, systems, processes, or practices that can be improved to enhance productivity and the financial bottom line. There are four critical areas where attention must be paid to prepare employees to become and

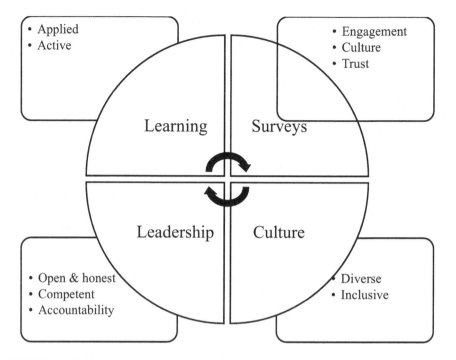

FIGURE 9.5 Engagement Tools for Readiness and Sustainability.

remain technically competent and highly engaged in the business of the organization. These areas (Figure 9.5) can be used as tools and the positive information or
experience gained can be used as best proactive to enhance an organizations performance and gain the commitment of employees.

9.5.1 LEARNING

There are many aspects of learning that have been deployed to prepare students and
professionals to be productive subject matter experts (SMEs). However, there are two
processes that are highly effective in helping an individual retain information and
put it into use. Active and applied learning processes are extremely important in the
development of technical workers such as design engineers to improve their critical
and strategic thinking capability.

The process of active learning involves having the student engaged in the process of
learning through reading, writing, discussions, and solving problems both complex and
simple. The process of active learning can be used to promote critical thinking. Some
active learning activities that can be used to facilitate critical thinking in the classroom
include (Kyoungna Kim, Priya Sharma, Susan M. Land, and Kevin P. Furlong, 2013)

- Using scenarios that will require students to apply their knowledge by providing a justification for their responses
- Introducing problem-solving activities that require students to engage in
  data analysis, evaluation, and synthesis

- Utilizing problems that have more than one possible answer requiring students to consider different alternatives and views
- Using open problems that require students to apply key technical concepts such as engineering, biology, chemistry, and physics to the solving problems encountered every day
- Utilizing open problem-solving in small groups and encouraging participation and input from all group members
- Using collaborative learning techniques to group projects that require the use of dialogue and social interaction in learning and conducting projects that require the capabilities and abilities of others to reach a decision or complete a task
- Writing essay and research papers

Applied learning is an approach that has been utilized through partnering and collaboration with corporations and educators and also used by corporations to enhance knowledge and skills of workers. This type of learning is an approach that actively engages students and professionals in directly applying the skills, theories, and models that they have learned through mostly classroom exposure. Applied learning may also be referred to as "real world work experience." It can be achieved through, for example,

- Work experience in the industry of interest
  - Can be gained through shadowing
  - Can be gained through internship
  - Can be gained through taking on assignments as a volunteer
- Taking on an assignment on a project team

9.5.2 Surveys

Conducting surveys continues to be the most popular means used by organizations to obtain feedback on the performance of various aspects of their environment. The advantage of using survey instruments to evaluate employee engagement, organizational trust, and culture is that the results obtained can serve as the basis for comparison and generalization of the data collected over time. For example, a baseline can be established using the data collected from the first time the survey is administered. The analyzed data can be used to serve as a baseline, and subsequent data collection activities could be compared to the baseline to determine if progress is being made. An improvement strategy should be developed based on the results of the data. The results of the data and the improvement strategy should be shared with organization members.

9.5.3 Culture

Culture is one of the most effective tools that can lead to engaged workers and their ability to be prepared to support the vision of an organization and sustain the ability to increase profitability through technological advancement, products, and services. Performance of humans in an organization is greatly dependent upon the culture of that organization. Culture is a powerful force that directs the life of individuals and bonds relationships in an organization. It is recognized that a trusting culture facilitates a learning culture where people are not afraid to take risks and allow creativity to flow in development of new technologies, products, or services that will improve performance of their organizations. Culture is the foundation in which trust develops and grow. Trust is credited with contributing to increasing positive workplace behaviors, such as positive attitudes and an increase level of performance. Trust in organizations is easier to achieve when the aim, vision, values, mission, and goals are shared, understood, and embraced (Alston, 2014).

It has been said that performance will improve in areas that are focused on; therefore, when organizational leaders focus on developing a culture of engagement, the result is that the organization has more engaged employees. Some attributes of an engaged culture include

- Providing advancement and learning opportunities
- Seeking feedback from workers on ways to solve problems
- Allowing workers to explore and exercise their creativity
- Embracing diversity and inclusion or workers

9.5.4 LEADERSHIP

Leadership plays a significant role in all organizational matters to include developing, nurturing, and creating a highly engage workforce. In order to have more engaged employees, leadership themselves must be viewed as actively engaged in the business and must focus on developing a culture that is supportive of involving engaged employees. Leadership must possess specific characteristics that are able to facilitate the success; however, there are three important aspects that must be at the top of the list. Among these characteristics are open and honest communication, accountability, and competency.

Leaders are incapable of performing their roles as leaders successfully if they are not viewed as being open and honest. Open and honest leaders effectively, timely completely, and accurately communicate information. They are not spinsters and only communicate the information that will be viewed as positive and withhold the information that may not be widely accepted or viewed as less than desirable.

Effective leadership provides them access to workers that are willing to go the extra mile to accomplish work. Effective leaders also seek to build trusting relationships with employees. As stated by Eric Shinseki, the former chief of staff of the army, "You must love those you lead before you can be an effective leader, you can certainly command without that sense of commitment, but you cannot lead without it. And without leadership, command is a hollow experience, a vacuum often filled with mistrust and arrogance."

Openness and honesty were identified in a study conducted by Alston (2014) as one of the most critical elements in developing trust, which is needed to be an effective leader. A second critical characteristic is accountability. Accountable leaders take responsibility for the performance of the organization, including the good and bad. Leaders that demonstrate accountability during the performance of their roles will incite others in the organization to take responsibility for their roles and actions and will reach a high level of accountability. The third critical characteristic for leaders is one of competency. When referring to leadership competency, reference is not made to the technical knowledge of the leader. A competent leader is competent and skilled as a leader and is able to motivate employees to follow their directions. Too often leaders are selected for a role in leadership because they have performed extremely well as a technical professional. In many cases these leaders require extensive leadership training in order to perform as leaders. Selecting competent leaders is extremely important for the success of organizations. Leadership can facilitate employee's readiness and sustainment by

- Encouraging employees to take an active role in organizational matters
- Providing opportunities for employees to grow personally and professionally
- Connecting employees to the vision and mission of the company
- Communicating the role of each employee in assisting the company to achieve goals
- Communicate the importance and role engagement play in the success of the company

9.6   APPLIED LEARNING

Case Study—review the case study and answer the questions below. In order to fully answer the questions, assumptions must m made. State any assumptions you make as a part of your response.

The Research and Development Corporation of America is an international company the design and manufacture various types of fitness products. The company chartered a team to begin working on a technology that will automatically design a fitness program tailored to each individual. The design team consisted of seven professionals, six males and one female. The team consists of three sedimentary workers that spent the last five years in the same department in the same position and four new design engineers who had been out of college for less than one year. Although considered to be great design engineers, the team members rarely network, explore new technologies or attend workshops or engage in activities that will ensure that they have current knowledge in their areas of expertise. The engineers with less than one-year experience felt that they had little input into product design and functionality.

The product was designed to be worn on the wrist for at least eight hours per day to provide accurate information that can be used to determine metabolic process and gather data that can be used to tailor a fitness program for the wearer. During design, the assumption was made that the users will be alerted not to expose it to water because it was discovered that water can impact the accuracy of the device.

1. What potential biases were introduced in the scenario that could impact the product development process and success of usage?
2. What role does critical thinking play in designing the fitness device?
3. What role does strategic thinking play in designing the fitness device?

4. Discuss the role of bias understanding in product development and its impact. What biases may have impacted the product design and the effectiveness of use?
5. What role, if any, a gap in knowledge may play in designing the fitness device? Explain your answer.

Additional Questions

1. What are the three primary means of developing a reflective practitioner in organizations? Discuss each and its importance.
2. List and discuss at least five soft skills that are important to technical professionals to ensure success in their roles.
3. Discuss strategic thinking and the role it plays in the engineering design and technology development process. Include in your discussion the aspects of strategic thinking and the actions that demonstrate that critical thinking has taken place.

BIBLIOGRAPHY

Allen Patricia Melton, Alston Frances E., Dekerchove Emily Millikin, *Peak Performance How to Achieve and Sustain Excellence in Operations Management*, CRC Press, 2019

Alston Frances, *Culture and Trust in Technology-Driven Organizations*, CRC Press Taylor & Francis Group, 2014

Alston Frances, *Lean Implementation Applications and Hidden Costs*, CRC Press, 2017

Belousova Alla, Pishchik Vlada, Technique of Thinking Style Evaluation, *International Journal of Cognitive Research in Science, Engineering and Education*, Vol. 3, No. 2, 2015

Brookfield, Stephen D., *Teaching for Critical Thinking, Tools, and Techniques to Help Students Question Their Assumptions*, John Wiley & Sons, 2012

Kallet Michael, *Think Smarter Critical Thinking to Improve Problem-Solving and Decision-Making Skills*, Wiley, 2014

Rutherford Albert, *Elements of Critical Thinking, A Fundamental Guide to Effective Decision Making, Deep Analysis, Intelligent Reasoning, and Independent Thinking*, Kindle Direct Publishing, 2018

Kyoungna Kim, Priya Sharma, Susan M. Land, Kevin P. Furlong, Effects of Active Learning on Enhancing Student Critical Thinking in an Undergraduate General Science Course, *Innovative Higher Education*, Vol. 38, 2013, 223–235. doi: 10.1007/s10755–012–9236-x

10 Future Directions
Towards the Humane Technology Designer

10.1 INTRODUCTION

Technology growth continues to be at the center of the economy in terms of providing profits for businesses, influencing the many factors of daily life for users, and enhancing the quality of life for the population that can take advantage of the various technologies. Being able to take advantage of a specific technology and utilize it to the fullest potential cannot always be achieved by everyone for reasons such as

- The technology may not be designed in a manner that can be used by a diverse population.
- The technology does in some way create a hardship for a particular group of users.
- All users cannot use the technology equally to the fullest.

Achieving the goal of developing products that can be used by a diverse population with a diverse culture is not always easy. Past experiences have revealed the difficulty of designing products and technologies for a diverse multigenerational population, as inconsistencies in product use has occurred and different groups of people have been impacted. The goal moving forward in new product design should be to minimize or completely eradicate the inequity and inconsistences in the ability for usage by a diverse populace.

One of the first steps required to design, build, and implement technologies that can be utilized by a diverse populace is to recognize the needs of every person and not just a target population. Followed by ensuring designers are trained and have the knowledge needed to design the desired products, understand that not all potential users are the same and that there are some key differences between the various groups of people, and gain an understanding of the customer base to inform new product design decisions.

10.2 THE HUMANISTIC TECHNOLOGIST: PRACTICE IMPLICATIONS

People who are knowledgeable and have the appropriate skills and experience are behind new product design, the workings of technology, and overall process development. The knowledge acquired includes a good foundation in areas such as

DOI: 10.1201/9780367854720-12

- Computer-aided design (CAD) and other design software
- The ability to synthesize information and make decisions
- The capacity and capability to think outside of the box
- The knowledge and needs of the customers that will be utilizing the end product

A humanistic technologist is defined as an individual engaging in technical design decisions that focus on ensuring the decisions yield products which are equitable, culturally calibrated, and do not create negative consequences to diverse unique populations. In today's world, the role of a humanistic practitioner is not actively defined and considered when educating engineers and technical professionals. As such, extensive considerations for the population of use, and their potential differences, are not considered as prominently as should be to ensure diverse usability.

There have been some realized implications for not effectively utilizing considerations for the human element during product design and development. There have been documented accounts of products not meeting the needs of specific groups of people within the United States. Some of those accounts have been introduced in this book to demonstrate specific points. The role of a humanistic technologist is expected to close the gap between the system approach and technology design.

Consider the implication of having a team trained in humanistic technology or just having a humanistic technologist on the design team functioning as a subject matter expert (SME). Implementing the design process utilizing the humanistic approach is expected to add value and sustainability for products produced for a company while creating equal opportunity for use by customers. Products and technologies designed with the assistance of a human technologist should possess, at a minimum, the following characteristics:

- Design with the diversity of users in mind
- Products are capable of being used without creating negative consequences to any one group of users.
- Products are designed in a way that will facilitate equitable use by the population.
- Product is not designed to benefit any one group.
- Cultural and sociological impacts are considered for real time product roll-out and into future usage.

The use of individuals that are knowledgeable or specialized in the area of humanistic technology can be considered a sustainable advantage for a company when the desire is to

1. Limit products which can have a negative impact on users and that can result in legal expenses for the company and the users (as in the AI technology example discussed later in this chapter)
2. Minimize the cost and potential stigma associated with pulling products from the market
3. Minimize expenses associated with retrofitting or redesigning products to enable usage

These are just a few reasons why retaining a skilled humanistic technologist on staff may provide a sustainable advantage for a company.

10.3 IMPLEMENTATION OF HUMANISTIC SYSTEM THINKING FOR PRACTITIONERS

The current system of thinking represents a comprehensive approach to analyze and take into consideration all parts of a system while focusing on the interrelationships and impact of each part on the entire system over time. **The discipline of humanistic psychology recognizes that humans are individuals who have unique differences. These differences must be recognized, evaluated, and considered during design-ing and developing systems and processes to add value to product sustainability.** Merging the belief of humanistic psychology and systems thinking forms the basis of a humanistic systems method of thinking. This way of thinking and decision-mak-ing is not currently practiced or is not practiced to its fullest. Systems thinking is the basis or underlying foundation for approaching design, coupled with humanistic thinking. Systems thinking can provide the following advantages for designs, the product produces, and the end users:

- Facilitates understanding how one part of a process or system influences another within the entire system, and the impact of the system on the human element
- Designs of products and systems will be a more equally distributed opportunity
- Reduces or eliminates human error during usage
- Increases product sustainability while eliminating or reducing the need to pull the product from the market for modification in the future
- Increases and expands corporate profitability and reputation

Humanistic systems thinking encompasses the attributes of both systems think-ing and humanistic design. Simplistically stated, humanistic system thinking is systems thinking with the human element included to obtain optimal usage. This approach to product design offers the most comprehensive approach to produce prod-ucts that are sustainable and provide value to a diverse customer base.

10.4 THE HUMANISTIC TECHNOLOGIST: ACADEMIC IMPLICATIONS

The path forward to produce highly skilled design engineers, who are capable of designing products that can be used by a diverse population, begins with training in the classroom. Classroom training for college students remains traditional in nature, meaning it currently does not focus on providing training such as on inclusion in design, culture, diversity, and inclusion. These types of training topics can provide engineers with a level of knowledge that can be built upon once they enter the profes-sional workforce. Currently, many technical workers enter the workforce without the soft skills needed to achieve success; therefore many companies impose additional training for entry level engineers to complement the technical training received in colleges and universities to assist in ensuring their success in the workplace culture. As such, **it is pertinent that colleges and educational institutions design curriculums that will expose students to the skills that will lead to products that can be utilized by people from all walks of life and facilitate success with the people aspects of their roles in industry.** People from all walks of life include, culture, race, sex, ethnicity, handicaps, body structure (short, tall, etc.). Developing technologies to suit such a large group of people is not easy, but every attempt must be made to achieve that goal.

Current curriculum and training presented in educating engineers is focused on the technical aspects of the specific engineering field, such as physics, fluid dynamics, and statistics. There should be careful considerations to the infusion of the humanistic aspect of technology design and application to ensure that products or technologies are applicable across the board.

The course curriculum listed in Table 10.1 or equivalent courses should be considered for inclusion in the curriculum for a typical engineering degree program. The table is an example of how to incorporate courses that will broaden a student's perspective on including the humanistic and environmental aspects into product and technology design. The example in Table 10.1 can be tailored for any engineering or technical degree program. Table 10.2 offers an example of how and where these types of courses can be inserted into a four-year college curriculum for engineers.

TABLE 10.1

Considerations for Changes to Current Curriculum for Engineers

Course	Potential Course Content	Course Purpose
Implicit Biases	Define implicit biases, types of implicit biases, ways to disrupt racial biases, how these biases impact the workplace, product, and technology design.	Provide engineers with an understanding of the impact of bias on decision-making; provide an understanding of how biases are formed and how to recognize and deal with bias as they occur.
Culture and Its Impact on Technology Development	Different culture and how they differ, technology view by different culture and affinity for usage, ability, and willingness to change and embrace new technology	Provide an understanding for designers on the population that their design will reach along with the cultural differences and how their technology will be used and if their technology will be used by the populace.
Cultural Diversity	A study into different cultures and their differences that can impact product design, development, and implementation.	Knowledge of the different cultures
Inclusion by Design: Future Thinking Approaches to New Product Development	Define and discuss the problem faced concerning inclusion during design.	To fill the gap in current teachings in colleges and universities. Includes future design thinking approaches.

TABLE 10.2

Example Future Electrical Engineer Curriculum

Freshman

Fall Spring

Engineering Physics I Engineering Physics II

Implicit Bias Engineering Introduction

English Composition I English Composition II

Calculus I Calculus II

Elective Computing for Engineers

Sophomore

Fall Spring

Differential Equations Algebra—Linear

Electrical Circuits Electronics

Science Computer Systems

Digital Logic Circuits Signals & Systems

Electrical Engineering Lab **Cultural Diversity**

Junior

Fall Spring

Chemistry Applied Electromagnetics

Chemistry Lab **Scenario Planning**

History Electric System Design Lab

Electric Power Engineering

Culture & Technology Development

Senior

Fall Spring

Inclusion by Design Senior Design Project II

Ethics Elective

Senior Deign Project I Thermodynamics and Fluids

Engineering Economics Engineering Mechanics

Elective EE Seminar

10.5 IMPLEMENTATION OF HUMANISTIC SYSTEMS THINKING FOR EDUCATORS

Implementing humanistic systems thinking may be a little complex as many professionals are accustomed to utilizing a general systematic approach in completing tasks and projects. The systematic approach has been around for decades and has been refined into what some people consider as the most effective and efficient means of working within a system. Introducing the humanistic approach will undoubtedly introduce complications in the thinking process for many professionals. Although complex, there must be a concerted effort to overcome these complications. A strategy to overcome complications when incorporating humanistic considerations includes the following:

- Training on various disciplines and topics such as
  - Humanistic approaches and considerations
  - Explicit biases
  - Diversity and inclusion
- Information sharing and feedback sessions
- Design collaboration and research
- Team-based approach to problem-solving

10.6 THE HUMANISTIC TECHNOLOGIST: PATHWAY FORWARD

Recall the three tenets of inclusive design that were introduced and discussed in detail in Chapters 2: consequential, equitable, and culturally calibrated. These three tenes are instrumental in contemplating how to implement and ensure that inclusive design concepts are embedded in the process and practices used in designing products. Systems and technologies are to be utilized by a contingent of users from differing cultures. The humanistic technologist or practitioners should be equipped with the tools necessary to design for present-day and future technologies. Designing for the humanistic interface touching many aspects of life has been gaining traction due to the increase in knowledge and exposure to failures that have created adverse situations resulting from technologies that have negatively impacted individuals and groups. Sections 10.7 through 10.9 provide detailed information of implementing and integrating the three concepts into your processes and practices.

10.7 CONCEPT OF EQUITABLE DESIGN

Equitable design produces products, services, and technologies that are useful to the marketplace, can be used with the same level of ease by consumers with diverse abilities from diverse backgrounds and culture, and provides the same use opportunity for all users. **An important aspect of the ability to design with equitability is to invoke the human element of empathy to position oneself to be able to see things through the lens of others that are from differing backgrounds and have differeing needs.** Looking through a design lens that is not traditional to the way an individual would normally view things and think can bring into consideration the needs of others and the way others will use and interact with the product. For many, this may be easier said than done; however, concentrated efforts must be made to accommodate and include a diverse population during design planning. Although not included in the key attributes, the importance of empathy in preparing to focus on the equitable concept and what is needed to move forward with equitable purpose should not be underestimated. In fact, empathy is key in successfully maneuvering through each attributes of this concept.

The tenet of equitable design has three critical elements that are important to render a product, technology, or service as being equitable for all potential users. These elements and associated attributes and are shown in Figure 10.1. Each of these attributes forms the basis of designing a product or technology that can equally serve the needs of the populas regardles of race, sex, cultural affinity, physical capability, or gender. The focal point is that the product or technology does not underserve any particluar group of the populace or does not provide an enhanced value or favor for

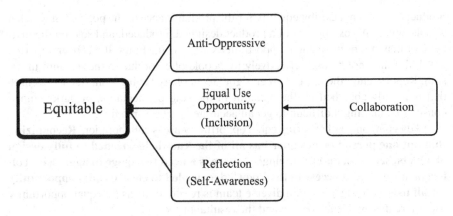

FIGURE 10.1 Key Attributes of the Equitable Tenet.

any one group. **Designing with equity is a deliberate and focused action toward an attempt to achieve an equitable product**.

10.7.1 ANTI-OPPRESSIVE

When designing technology, it is advantageous to ensure that the design does not create oppression for any one group of users. Oppressive products or technologies are capable of inflicting hardship and constraints on some users. However, if just one group is negatively impacted and leads to oppression, the technology is considered oppressive. **When referring to anti-oppressive design, reference is being made to designs that produce products and technologies that do not create hardship or disempower a particular group of people**. For example, products or technologies that cannot be used by a particular group of people because it does not produce the same impact, benefits, or effect as it would in other groups or parts of the population.

To give you an idea of what is referred to as anti-oppresive design, refer to the example of an oppressive design in an article written in the *New York Times* titled "Wrongfully Accused by an Algorithm" (www.nytimes.com/2020/06/24/technology/facial-recognition-arrest.html). As previously discussed, the article documented flaws in the design of an artificial intelligence system used by law enforcement. As a result, an innocent man was falsely accused and arrested for a crime that he did not commit. This false accusation resulted from the use of a faulty AI technology. An extreme hardship was created for the individual as a result of the false imprisonment. Consider the trauma to the family, the legal expenses paid, embarrassment of being labeled a criminal, and the list goes on. Eventually the issue was resolved, but the lasting impact created by a flawed technology will not be forgotten by the man and his family.

10.7.2 EQUAL OPPORTUNITY USE

The yield of the optimal designed technology, product, or system will afford equal use opportunity for the preponderance of the populus. Throughout this book, we provided several examples of equal opportunity use for various technologies and products that did not exhibit equality for the prodominance of the populus in which it was designed. An example of of a product designed, produced, and sold on the market that may not provide equal opportunity for use is the Fitbit. It has been reported that Fitbit may not be used effectively by people of color due to the melanin in the skin. Reports cited the use of "green light" technology having a shorter wavelength that is readily absorbed by the melanin in the skin pigmentation of dark-skinned consumers, making it difficult to get an accurate heart rate reading.

A key component of ensuring equal opportunity use is collaboration. **Recognizing that not one person or designer has all of the knowledge needed to fully design products, services, and technologies, there is a need to engage in aggressive collorabration to gain access to the knowledge needed to ensure equal opportunity for all users.** Engagement of a diverse team is required to ensure equal opportunity use is a reality with each design and the resultant products.

10.7.3 REFLECTION

Mindfullness sets the stage for better focus, which is paramount in technology and product development, as the lack of focus can introduce errors that can follow the design through development and on to customer usage. It provides an opportunity for the design team to think about users and the evolution of the product or technology and its impact over time. Mindful design simply is spending time concentrating on and purposedly paying attention to the design process, outcome, and end users. Mindfullness is a key component of the reflection process for inclusive design. **Reflection represents a deep look into the decisions made, the thoughts encountered, and critical decisions acted upon.** Considering the previously mentioned Fitbit example, reflection may have assisted in resolving the issue of technology selection in deciding to use green technology alone, instead of other available more capable technology.

10.8 MOVING TOWARDS A MORE EQUITABLE DESIGN

There are some key considerations for leadership and teams listed in Table 10.3 to help close the gap in creating designs that can be considered to be equitable for all users. This table is not an all inclusive representation of options to be considered; however, the table is intended to place ones' thoughts on the path that could assist in creating equitable designs.

10.9 CULTURALLY CALIBRATED

The United States (US) is representative of an heterogenous culture because people from many different countries have settled in the US, with many of them becoming citizens. Becoming a citizen of the US or migrating to the US does not mean that the culture that the immigrant came from was left behind. In fact, one can be assured that the former cultures accompanied the immigrant, thereby integrating their culture into society in some facet. Having this knowledge of the makeup of the US alone provides knowledge that a designer can use to ensure that culture is at the

TABLE 10.3
Actions to Close the Gap in Developing Equitable Designs

1. Engage a diverse team that includes potential underserved populations
2. Provide training to team on diversity and inclusion and cultural impact on designs
3. Discuss team dynamics and the role of collaboration and idea sharing
4. Discuss and demonstrate leadership modelling the attitude and behavior desired from the team
5. Administer team sensitivity briefings/training
6. Schedule reflection time for the team together as well as individually
7. Provide a reflective logbook and requires documenting the process and reasoning behind decisions8. Encourage open and honest communication (positive and constructive) and the free expression of ideas
9. Assemble a diverse user group and seek feedback on the design applicability and usage. Gather information of potential issues that will impact usage. Ensure that underserved groups are represented as an active part of the team
10. Expose team members to situations where products have worked for one segment of the population and discuss issues, short commings, and solutions
11. Ensure that the team is aware of the negative impacts of culture and consider appropriate changes to design where appropriate

forefront of decisions made during the design process. **Culture has a significant influence on what is viewed as an acceptable design, unacceptable design, a bad design, or a good design by members of a specific population.** Developing a business culture that supports exponential growth can be challenging because of the diversity of people and the different values, practices, beliefs, and behavior they bring to organizations. These cultures, along with the varying subcultures, form the basis of the underlying overall culture of an organization. Although challenging, it is incumbent upon leaders to leverage these differences into business ventures so design teams can develop the best products that enhance inclusivity. In addition, a learning culture among organizational members and design teams is necessary to embrace the diversity of cultures that each individual brings to the organization and team so that inclusivity can take root. Organizations having a learning culture facilitate continual improvement, strategic thinking, and innovative practices while reflecting on and learning from past mistakes. **Culture can place substantial influence on an organization or team because the shared values and beliefs that are present within a culture represent the variables that guide behaviors** (Alston, 2014).

Cultural calibration is an important aspect of designing for inclusivity because it provides focus on the users from a global perspective. Recognizing that different cultures will inevitably have different values, practices, beliefs, and exhibit differing behaviors will add clarity and direction for product and technology design. The key attributes of the culturally calibrated concept are shown in Figure 10.2. A culturally calibrated organization and design team understands the importance of these attributes, the role culture plays in the final product produced, and the impact of design by a diverse knowledgeable team and orgnaization.

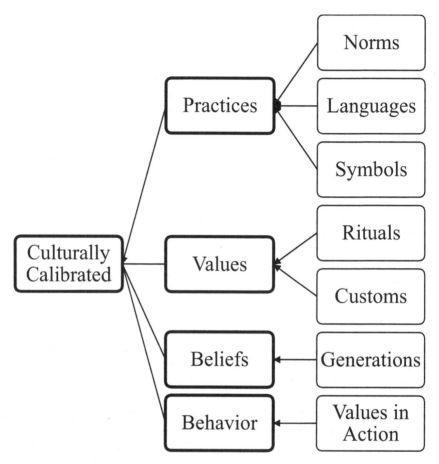

FIGURE 10.2 Key Attributes of The Culturally Calibrated Tenet.

10.9.1 PRACTICES

When speaking of the practices within a culture, reference is made primarily to the traditions, norms, and languages of people within that culture. These practices may be and are often different from other groups and are manifested by way of the norms which include the shared expectations that are seen: spoken language and prevailing symbols visible for all to see. **Norms are standards and nonverbal unwritten rules that people live by.** Consideration for the practices that are prevalent in a culture should be considered in the design phase of products and technology because it can make a difference whether the product can or will be used by the populace. For example, in the US tattoos have been gaining prevalence so much so that many people have tattoos across their body, and in some cases, it is not unusual for body parts to be almost completely covered with tattoos. It has been reported that tattoos can be an impediment to the Fitbit operational accuracy because the ink used to apply them creates a darkening of the skin. Having knowledge of the practices that are prevalent within a culture, a design can be constructed to complement the end user's cultural integration needs.

In addition to focusing on the culture of future customers, it is just as important to focus on the practices within the design team. In fact, focusing on the practices of the design team should be first and foremost in the minds of the leadership team and the team leader. Team practices should include engagement of all members, extending a listening ear without prejudgment, recognizing that every thought and opinion is valuable to team outcomes, embracing the differences in cultural contribution and differences of each member, and approaching all decisions without bias to ensure conclusions are based on sound data.

10.9.2 VALUES

Values are important aspects of culture because they tend to determine the current and future behaviors of members. Values are important to ascertain for the team as well as for the population in which the design is slated to attract for usage. **With regards to the design team, it is necessary for the team to value listening and being mindful of the information content without a rush to judgement, making decisions based on nonbiased information and consideration for the customer and the impact of new products and sharing knowledge and experience so that the entire team can benefit and grow.** It is paramount for the design team to understand the value system of the population for which a product or technology is being designed.

10.9.3 BELIEFS

Beliefs and values are closely aligned, although beliefs are generally passed down from generation to generation. In product design, one's belief is important because it forms the attitude as to whether something is true or false. Furthermore, it forms the assumptions that we make about people and our expectations of them; for example, the way they act and live. Keep in mind that our beliefs of others are not always accurate, and in a team setting it is important to understand each team member and be sensitive to their beliefs because they may, in fact, be different. Engrained beliefs are not easy to change; however, beliefs can be understood and used in a positive manner to enhance team cohesion and productivity. Beliefs should also be considered when designing products that will reach a diverse population.

10.9.4 BEHAVIORS

Behavior consists of what a person does; behavior is readily observable and can serve as a view into an individual's value system. Because behavior is visible, it is easier to observe and determine a solution for correction and modification. It is important for an individual to have the ability to self-direct and modify practices with relative ease. Team behavior can make or break a team's ability to attain cohesion as well as design products and technologies to benefit prospective users. Destructive behaviors by team members include impatient listening, reserved or incomplete communication, lack of engagement, myopic way of looking at things, and lack of focus and self-awareness.

Just as there are team behaviors that are destructive, there are desired behaviors for team members that facilitate creativity and productivity. **Desired behaviors include patient listening skills, active communication, respect for other team members, trust extended to team members, accountability for the work in which they have been asked to deliver, demonstrating respect for each other, and dedication the team.** Finally, remember that a team will fail or succeed together, and the design of the resultant product is the shared responsibility of the team.

10.10 TOWARD A MORE CULTURALLY CALIBRATED DESIGN

Culture is a key consideration for product or technology design. It is costly for a company to design, manufacture, and distribute products that can only be used effectively by a small population because of cultural differences. Therefore, cultural calibration is a concept that each designer should have in mind when embarking upon the design process. In order to close the gap on creating designs that take into considerations the impact of culture for all users, there are some key considerations for the leadership and teams listed in Table 10.4.

10.11 CONSEQUENTIAL

The consequences that can be experienced from a flawed design that does not take into consideration the concepts of inclusive design can range from minor to catastrophic impacts to the company as well as their clients. A good example of this type of consequence was documented in an article written in the *New York Times* titled "Wrongfully Accused by an Algorithm." The article documents the flaws in the design of an AI system used by law enforcement. Earlier studies by M.I.T. and the National Institute of Standards and Technology, found that the technology employed by these systems worked well on white skin males and was less accurate for darker skinned people because of a lack of diversity in images used in developing the product and its systems. Because of the failed AI system, an innocent person and his family had a terrifying experience with the legal system that they will never forget.

To avoid negative consequences to people with skin pigmentation, the engagement of strategic and critical thinking is necessary to assess the potential impacts to the

---

TABLE 10.4

Actions to Close the Gap in Developing Culturally Calibrated Designs

1. Engage a diverse team that includes potential underserved populations
2. Provide training to team on cultural impact on design
3. Leadership modelling the attitude and behavior desired from the team
4. Assemble a culturally diverse user group and seek feedback on the design applicability and usage. Gather information of potential issues that will impact usage due to cultural values, beliefs, and practices
5. Ensure that the team is aware of the negative impacts of culture and consider appropriate changes to design where appropriate

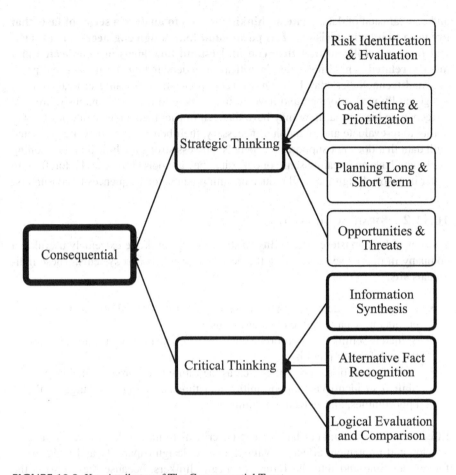

FIGURE 10.3 Key Attributes of The Consequential Tenet.

various people. Critical thinking and strategic thinking concepts are key in anticipating, recognizing, and controlling negative or unwanted consequences. Figure 10.3 shows the key attributes of the consequential tenet. These attributes when invoked and utilized effectively will certainly minimize and, in some instances, eliminate certain negative consequences of the product to the user and the company. The two attributes of critical and strategic thinking which serves as the foundation of everything a design professional does and is the foundation of the approach used to find solutions to problems, make decisions, develop new ideas and technologies and processes, and unleash their creativity.

10.11.1 CRITICAL THINKING

Critical thinkers are highly effective because they have the ability to optimize and synthesize information large or small and formulate a defensive conclusion, recognize and put into perspective alternative facts (untruths), and conduct the logical evaluation and comparasion of data. **Critical thinking is used to analyze a series of facts that are used to inform decisions. It is paramount that design engineers possess critical thinking skills so that they can understand how ideas are connected and are interelated.** This knowledge is critical in understanding the impact that products and technologies may have on a diverse population. Because critical thinkers are generally well-informed and have an inate desire to learn continuously, are able to recognize biases that they may have that can impact their judgement, possess the ability to re-evaluate and abandon, if necessary, their beliefs when they are presented with data that does not support their beliefs, possess strong analytical and reasoning skills, and have an uncanny sense of curiosity, they are poised to be at the forefront to foster inclusive design that will reduce or limit potential consequences to consumers.

10.11.2 Strategic Thinking

Those who have mastered the ability to strategically think are extremely useful to a company in many ways, including the design of products and systems. These individuals are

- Competent in identifying and evaluating the potential risk that a system or product may have on the company or user
- Skilled in setting the appropriate goals for a project or company and charting the path for completion
- Able to plan for the present as well as the future to ensure sustainability
- Skilled in identifying opportunities and threats, taking advantage of the opportunities, and reducing or eliminating the threats

Effective strategic thinkers have mastered critical thinking skills and are able to develop and implement effective plans to ensure design capability and inclusitivity for present time and into the future. Strategic thinkers, because they possess the skills of a critical thinker, also have the ability to recognize their biases and control the impact that these biases may have on their decisions and objectively review data.

10.12 TOWARD A LESS CONSEQUENTIAL DESIGN

As shown in Table 10.5 there are some ways to reduce or eliminate the impact of negative consequences to users when designing new technologies or services.

TABLE 10.5

Actions to Close the Gap on Negative Consequences within Design

1. Engage a diverse team focusing on underserved populations
2. Provide critical thinking skills training to team members
3. Provide strategic thinking skills training to team members
4. Ensure that there is an understanding of the cultural impact
5. Ensure that the team members include a segment of the population that is catergorized as underserved

---

10.13 APPLIED LEARNING

Technology will continue to grow at record pace. As such, designers must be equipped to meet the technological challenge while still executing the design process. To be equipped, changes in the way engineers are educated and trained is required, along with executing the design thinking process that produces inclusive designs, technologies, and products.

1. Define humanistic technologist. What role should this discipline play in the new product design process?
2. What changes or modification to the way engineers are trained would be appropriate to render designers ready and capable of designing inclusive products.
3. What are the three tenets of an inclusive design? Explain the role of each tenet in the design process. Research and provide a real-life example where failure to use each tenet causes complications for the users or the company.

BIBLIOGRAPHY

Alston, Frances, "Culture and Trust in Technology-Driven Organizations," CRC Press, Taylor and Francis, 2014.
NIST. www.nytimes.com/2019/12/19/technology/facial-recognition-bias.html
www.msn.com/en-us/news/technology/fitbits-might-not-track-your-heart-rate-right-if-youre-a-person-of-color/ar-AAERcrN; www.statnews.com/2019/07/24/fitbit-accuracy-dark-skin/
www.nytimes.com/2020/06/24/technology/facial-recognition-arrest.html
www.nytimes.com/2018/02/09/technology/facial-recognition-race-artificial-intelligence.html
www.theguardian.com/technology/2015/apr/30/how-tattoogate-has-highlighted-a-wider-problem-with-wearables

Part 3

Case Study

Grand XI Personal Fitness and Wellness Coach

11 Application of the 4D Model for New Product Design

11.1 GRAND XI COMPANY

The Grand XI Company is a multimillion-dollar conglomerate in the fitness and wellness business. Grand XI manufactures professional stationary bikes, elliptical machines, and treadmills and is considered an industry leader in integrating contemporary design into products that are functional, yet aesthetically pleasing and motivational to use. Their products are often used by interior designers in particular when designing home gyms. In addition to their fitness equipment product line, Grand XI has also developed a smart watch and program application that supports data collection from their fitness equipment. Grand XI's customer base is primarily in North America, including the United States, Canada, and Mexico, but they would like to extend their market share into Asia. The Grand XI Company has several product divisions and are looking for a way to globally expand their customer base through innovation and technology. In particular, the Grand XI Company has a vision of being the first fitness company to be a one-stop shop for improving the health and wellness of their customers. Customers could address all their wellness needs through Grand XI's technology. The Grand XI Company has decided to apply design lens and vision concepting, through the application of the 4-D Algorithm for New Product Design to design and deploy the next generation of wearable fitness technology.

11.2 GRAND XI COMPANY VISION AND MISSION

Leadership at the Grand XI Company has a vision that over the next 10 to 20 years there will be a desire to better manage the data that is currently being collected by traditional fitness watches and fitness equipment. Company leaders believe that the technology is common for collection of the health and fitness data, but how the data can be used and the process for applying the data, has not yet come to full maturity. In particular, as the COVID-19 pandemic has emerged, the Grand XI Company experienced unprecedented growth in customers purchasing their equipment and fitness programs for home gym use. The company recently entered into a partnership with a global non-profit organization focused on improving wellness of the overall individual, and the Grand XI Company is excited to be a part of an effort to improve the quality of life for people across the world.

Grand XI is looking to invest a significant amount of capital in research and development of an electronic program and wearable technology which can be used to

DOI: 10.1201/9780367854720-14

integrate data collected across all different types of Grand XI fitness equipment, including other wellness physiological data such as individual sleeping and activity patterns. The integrated data would then be evaluated by artificial intelligence (AI) to make fitness and wellness recommendations for improving each customer's life. Information from the non-profit organization would be incorporated into the program application in making wellness recommendations on topics such as diet, social interactions, spiritual reminders and plans, emotional guidance, and suggestions such as, "take 5 minutes to mentally focus with our Grand Yoga program." Recommendations for all aspects of an individual's lifestyle could be addressed: diet and exercise recommendations could be made, along with recommendations of when to consider seeing your physician for a checkup (e.g., not sleeping well with your sleep patterns identified). The overall device and associated electronic application program would function as a personal health and lifestyle coach. The Grand XI Company would integrate the app on the cell phone, computer, or any other smart device and would use wearable technology for data collection (ability to display the data across multiple electronic devices). The Grand XI Company is hesitant to further develop their fitness watch because the current market is saturated with similar products. Leadership at the Grand XI Company has asked their research and development department to lead this important research and development project in developing the next fitness product revolution and has decided to name the project *My Saturi*.

11.3   GRAND XI COMPANY OPERATIONAL INFORMATION

Total annual revenue of the Grand XI Company is approximately $675,000,000 and includes dedicated organization divisions for each product (e.g., Grand XI Stationary Bike). The Grand XI Company currently employs approximately ten thousand employees working at two locations within the United States. The organizational structure for the two locations is quite different and the company is considering applying and implementing the same organizational structure at both locations to improve management of the company but also to create consistency for employees who may work at either or both locations.

Over the past two years there has been a change in the senior leadership managing Grand XI; as such, the company has experienced an increase in complaints about the lack of opportunity for professional growth and lack of respect on a daily basis when interfacing with employees. A general theme of "productivity over safety" has emerged, and management is planning a series of initiatives targeting improving the overall culture of the company but is also considering how to effectively manage organizational culture issues across the different company divisions.

The organizational structure for one location is aligned with a traditional functional organization, and the organizational structure at the other location is aligned with the traditional matrix organization. Grand XI is planning to manage the research and development of the new Grand XI Fitness and Wellness Coach with a matrixed organizational structure. All design team members will be permanently assigned to their home organization but would temporarily be assigned to a project manager for developing the Grand XI Fitness Coach product.

TABLE 11.1
Grand XI Company EEO Report

Total	Males	Females	Other
	6809	3189	2
	Nationality		
White	4945	2042	1
African Americans	1032	753	0
Latino	600	351	0
Asian	120	40	0
Other	112	3	1

The Grand XI Company was founded in 1987, and since 2012, the company has targeted improving diversity and talent of the staff. Table 11.1 identifies results from the company's most recent equal employment opportunity (EEO) annual report conducted by an external consultant. The company has been focusing on adding more diverse employees to the staff; however, the progress has been a little slow in attracting the talent needed to further enhance the organization and its design and engineering capability.

11.4　GRAND XI COMPANY PROJECT MANAGER AND PROJECT PLAN FOR THE *MY SATURI PROJECT*

Art Awesome was selected as project manager by the company's chief executive officer to lead and manage the project due to his past experience and success in managing research and development projects for Grand XI. Art is experienced and has had extraordinary success in managing different project types and has consistently delivered projects on schedule and within budget. As the project manager, Art completed the following for the *My Saturi Project*:

- Development of the project plan
- Development of a resource loaded schedule
- Assembled the project team
- Coordinated with executive management and accounting on the project budget
- Ensured that team members were notified of their selection and role on the project
- Scheduled and conducted the project kickoff meeting with team members

11.4.1　DESIGN TEAM ASSEMBLY AND MEMBERSHIP

The project manager collaborated with the functional managers in selecting and appointing project team members. Table 11.2 lists the demographics of the employees that were chartered to work on the project team. The number of team members

TABLE 11.2

Project Staffing of *My Saturi* Project Team

	Design & Engineering	Business & Marketing	Finance	Purchasing	Production Start-up
Males	37	10	4	6	10
Females	15	5	2	4	7
Caucasian	37	14	5	5	12
Non-Caucasian	15	1	1	5	5

fluctuated during the various stages of the project. Team members ranged from 50–100 active members depending on the stage and progress of the project. Some team members were previously assigned to the other company location. In addition, some employees were subcontractors from a prominent engineering firm that the company had teamed up with on several other projects. The reassignment of personnel was completed to secure the talent needed for the project and to further diversify the project team.

11.5 STEP 1: DISCOVER—DATA COLLECTION FOR MY SATURI PROJECT

As depicted in Figure 11.1, the first step in applying the 4-D Algorithm for New Product Design is the **discover** step which includes the collection of data to be used by Art's team in the design process. Qualitative data collection methods were selected for gathering research data. The data collected was designed to support the application of vision concepting when implementing Step 2 of the model (Define), Change Factors and Selection of Drivers.

The design team evaluated the different approaches for collecting data and selected three types:

- Questionnaire survey. The survey targeted a large number of consumers to better understand how important health fitness and wellness is to their lives and how the data was being used.
- Focus groups. Focus groups were used to expand on information collected in the questionnaire survey and to support marketing strategies of the new fitness and wellness product.
- Market analysis and benchmarking against other fitness and wellness companies of how their data is collected and used

Art and his project team recognized the usefulness of the data collected must be directly relevant to the tools used for analyzing the data; therefore, data collected would need to support Step 2 of the 4-D Algorithm for New Product Design. Art assigned Susan White as the lead for collection of the data related to understanding the different types of wearable technology, people's perceptions of the technology and usability, and general information people are interested in when it comes to fitness and wellness. Susan and the project team identified design lens which were

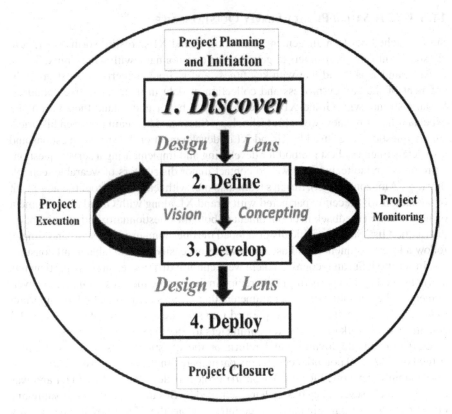

FIGURE 11.1 First Step of the 4-D Algorithm for New Product Design.

important to Grand XI's customer base and would need to be incorporated into the different approaches for gathering data. Listed below are several of the more important design lens identified and incorporated into the different data collection methods.

- *Trust in the data* generated and confidence in validity of the results. This data parameter was targeted to be collected when implementing the questionnaire survey and focus groups.
- *Health consciousness* of today's consumer. This data parameter will be collected when implementing the questionnaire survey and focus groups.
- *Cost acceptance.* This data parameter will be collected when implementing the questionnaire survey, focus groups, and market analysis.
- *Sustainability and durability of the technology.* This data will be collected when implementing the questionnaire survey and focus groups.
- *Comfort in wearing the technology.* This data will be collected when implementing the questionnaire survey and focus groups.
- *Security of the data.* This data will be collected when implementing the questionnaire survey and focus groups.
- *Regulatory and licensing requirements of the consumer's home country.* This data will be collected when implementing the market analysis.

11.5.1 *My Saturi* Project Survey Questionnaire

Susan reached out to management at existing Grand XI stores in North America to assist in identifying consumers targeted for the questionnaire within the United States, Canada, and Japan, and met with key fitness and wellness experts to assist in establishment of the survey process and collection of data in their respective countries. Susan also met with Grand XI key leadership to better understand their vision and expectations for the new wearable technology. Susan and her team reviewed historical survey questionnaires used by Grand XI in development of the survey questions and also determined the best method for developing and implementing a survey questionnaire on wearable technology was to contact major distributors of wearable technology (e.g., Amazon, Target). Susan's team also met with leadership from the non-profit organization who recently partnered with Grand XI, along with Grand XI equipment owners to solicit feedback in development of both the questionnaire and questionnaire distribution list. Susan specifically developed the questionnaire to be short and simple, follow a logical sequence, and use common terms easily recognizable to all consumers. In addition, Susan recognized there were cultural differences in study participants and consulted with experts from each region in developing the questionnaire. In development of the questionnaire, the parameters that Susan's team identified as important to Grand XI customers were incorporated into specific surveys questions. Table 11.3 contains a crosswalk of several key parameters to specific survey questions.

Development, deployment, and return of the questionnaires took six months. A total of 5,478,310 potential consumers participated in the survey. An excerpt of the questionnaire developed for the *My Saturi* Project is depicted in Table 11.4 and was cross walked to several key design lens identified by Susan and her team. A summary of the data for all participants (and countries) from the *My Saturi* Project Survey Questionnaire for several key questions are summarized in Table 11.5. Results of the survey for each country are further described below.

11.5.1.1 Survey Results for the United States

A total of 3,615,840 consumers participated in the survey for the United States. Analysis of the data collected from the United States resulted in the following conclusions:

TABLE 11.3

Crosswalk of *My Saturi* Project Key Parameters against Survey Questions

Key Parameter	Survey Question No.
Trust in the data	5
Health conscious	1, 2, 12
Cost acceptance	14
Sustainability and durability of the technology	8, 9
Comfort in wearing the technology	11
Security of the data	6
Ease of use	15
Reliability	8

TABLE 11.4

Excerpt of the *My Saturi* Project Questionnaire

Survey Question No.	Survey Question	Yes	No
1	Do you routinely monitor your health and wellness? [Health Consciousness]		
2	Are you interested in understanding any one of the following physiological parameters for yourself? Heart rate Blood oxygen levels Blood pressure Sleep patterns Respiratory rate Glucose level [Health Consciousness]		
3	Do you or have you ever used wearable technology such as a fitness watch?		
4	If you have ever used wearable technology, did you regularly use the information to improve your personal health and wellness?		
5	Do you trust the data you obtain from your wearable technology? [Data Trust]		
6	Do you have any concerns about how the health data generated from wearable technology could be used? [Data Security]		
7	Was the technology worn on the wrist?		
8	Did the technology maintain a charge as long as desired and adequately perform in different work environments? [Sustainable and Durable, Reliable]		
9	Was the material the wearable technology was constructed of durable? [Sustainable and Durable]		
10	Did the wearable technology interfere with any activity?		
11	Was the wearable technology comfortable? [Comfort]		
12	Have you ever used a fitness or wellness coach? [Health Consciousness]		
13	Do you want to improve your fitness and mental health?		
14	Would you spend $300 to have guaranteed results to improve your health? [Cost]		
15	Do you require technology that is easy to use and understand?		

- There is not a majority of people in the United States who monitor their health and wellness (and use fitness and wellness coaches); however, a significant number of people in the United States are interested in better understanding their physiological health.
- A third of the people in the United States have used wearable technology. When wearable technology was used, a significant number of people regularly incorporated the fitness information into their daily life.

TABLE 11.5

Summary of Questionnaire Survey Results by Country

Question No.	United States		Canada		Japan	
	Yes	No	Yes	No	Yes	No
1. Do you routinely monitor your health and wellness?	45%	55%	52%	48%	67%	33%
2. Are you interested in understanding physiological parameters?	67%	33%	70%	30%	75%	25%
3. Do you, or have you ever used, wearable technology, such as a fitness watch?	38%	62%	42%	58%	49%	51%
4. If you have ever used wearable technology, did you regularly use the information to improve your personal health and wellness?	73%	27%	65%	35%	82%	18%
5. Do you trust the data you obtain from your wearable technology?	75%	25%	78%	22%	83%	17%
6. Do you have any concerns about how the health data generated from wearable technology could be used?	52%	48%	35%	65%	7%	93%
7. Was the technology worn on the wrist?	95%	5%	92%	8%	86%	14%
8. Did the technology maintain a charge as long as desired and adequately perform in different work environments?	86%	14%	87%	13%	92%	7%
9. Was the material the wearable technology was constructed of durable?	95%	5%	96%	4%	94%	6%
10. Did the wearable technology interfere with any activity?	-	100%	-	100%	-	100%
11. Was the wearable technology comfortable?	75%	25%	78%	23%	73%	32%
12. Have you ever used a fitness or wellness coach?	15%	85%	22%	78%	10%	90%
13. Do you want to improve your fitness and mental health?	85%	15%	83%	17%	95%	5%
14. Would you spend $300 to have guaranteed results to improve your health?	90%	10%	92%	8%	93%	7%
15. Was the technology easy to use and understand?	80%	20%	85%	15%	90%	10%

- A significant number of people in the United States trust data generated from wearable technology and are somewhat concerned about how the data will be used and managed/secured.
- Almost all the wearable technology used in the United States was worn on the wrist; however, people would like the technology to sustain a longer charge and be waterproof.
- Wearable technology used in the United States is somewhat comfortable, easy to understand and use, and affordable at $300 or less.

11.5.1.2 Survey Results for Canada

A total of 492,893 consumers participated in the survey for Canada. Analysis of the data collected from Canada resulted in the following conclusions:

- Almost half of Canadians monitor their health and wellness and a significant number of people in Canada are interested in better understanding their physiological health parameters.

- Almost half of Canadians have used wearable technology. When wearable technology was used, a significant number of people regularly incorporated the fitness information into their daily life.
- A significant number of people in Canada trust data generated from wearable technology and are somewhat concerned about how the data will be used, managed, and secured.
- Almost all the wearable technology used in Canada was worn on the wrist, and most people were satisfied with the construction material.
- Wearable technology used in Canada was considered easy to use and understand and affordable at $300 or less.

11.5.1.3 Survey Results for Japan

A total of 1,369,577 consumers participated in the survey for Japan. Analysis of the data collected from Japan resulted in the following conclusions:

- There are a significant number of people in Japan who monitor their health and wellness and a significant number of people in Japan are interested in better understanding their physiological health parameters.
- Almost half of the people in Japan have used wearable technology. When wearable technology was used, a significant number of people regularly incorporated the fitness information into their daily life.
- A significant number of people in Japan trust data generated from wearable technology and are not concerned about how the data will be used and managed/secured.
- A significant amount of wearable technology used in Japan was worn on the wrist and people were satisfied with durability and sustainability of the technology.
- The majority of people in Japan have not used a wellness or fitness coach, but a significant number of people would like to improve their fitness and health.
- Wearable technology used in Japan is considered easy to use and understand, and affordable at $300 or less.

11.5.2 *MY SATURI* PROJECT FOCUS GROUPS

Susan and the project team analyzed the data from the questionnaire survey and selected a subset of questions, which would further clarify and provide additional information that could be used in the design process. When developing the focus group questions, the project team incorporated several key design lenses, similar to the process used in developing the survey questionnaire. Table 11.6 contains a crosswalk of design lens to specific focus group questions. Susan's team expanded on survey questions where additional information was needed to support analysis of the data, Step 2 of the 4-D Algorithm for New Product Design.

Susan recognized that the focus group questions may vary, by country, based on responses from the survey questionnaire.

TABLE 11.6

Crosswalk of *My Saturi* Project Key Parameters against Focus Group Questions

Key Parameter	Focus Group Question
Trust in the data	5, 7
Health conscious	1, 9, 8
Cost acceptance	13
Sustainability durability and reliability of the technology	14
Comfort in wearing the technology	2, 10, 11
Security of the data	5
Ease of use	2, 6

An excerpt from one of the focus groups questions lists is presented in Figure 11.2. Implementation of the focus groups included a strategy to incorporate diversity into the focus group mix.

Excerpt-Grand XI Focus Group *My Saturi* Project

1. Why is fitness and health wellness important in your life? Why or why not?
2. Do you or did you enjoy wearing the fitness technology? If not, why?
3. Do you feel like the technology is or was useful in better understanding and managing your health? Why?
4. What kind of information from the wearable technology is most useful to your fitness goals, and do you have concerns about how the data will be used?
5. Are you concerned about how the data will be stored, and are you concerned about any privacy issues associated with data management?
6. Have you ever used a fitness or wellness coach? If yes, were they effective in improving your life? Why not? Was cost a factor? Ease of use?
7. Would you use a fitness brand that could integrate all your fitness and wellness data to better improve your life?
8. Is there any reason in your life why you would not want to improve your health?
9. Would you want to learn more about your family's health and wellness? If so, what types of information would be useful to you as an individual and/or parent?
10. Would you be willing to wear technology on your finger or around your neck? Why or why not?
11. Do you have a preference on where the technology would be worn?
12. Would you be willing to wear the technology in all types of environments?
13. What is a reasonable cost for wearable technology?
14. How important is product reliability, sustainability, and durability?

Figure 11.2 Excerpt from Grand XI Wearable Fitness Technology Focus Group Questions

Summary of Focus Group Results

Results of the focus groups generated the following information:

- The majority of focus group participants identified that fitness and overall health and wellness was important to their life goals. The reasons for why fitness and health wellness were important to fitness group participants included overall improvement in their physical health, to address a health disease (i.e., rheumatoid arthritis), prolong their health and longevity, and maintain a healthy weight (appearance).
- 75% of the focus group participants indicated they routinely monitor the health of themselves and loved ones.
- 55% of the focus group participants said they own fitness-related wearable technology. Almost all the technology is related to being worn on the wrist.
- Only 7% of the group participants indicated that they were not interested in a wearable technology that is presented in the form of a necklace.
- Of the 55% of focus group participants who own fitness-related wearable technology, 65% of participants indicated they did not like wearing technology around their wrist and stated the location of the technology often interferes with how they perform their work. Although the technology can function as a watch, they do not frequently use the watch function because they have other devices they use for simple information inquiries (e.g., cell phone). In addition, personnel complained of having to charge the wearable technology daily and often do not wear it at night so as to not drain the battery, which defeated wearing the technology to understand sleep patterns.
- All but 2% of focus group participants felt like technology could be used to improve their fitness and wellness. Feedback was that participants desired the technology to be as easy as possible to operate and also desired a higher battery storage capacity.
- Focus group participants identified information listed in Table 11.5 as being important to their personal fitness goals. As noted, the information is primarily physiologically driven.
- Only 15% of participants stated they had or currently use a fitness coach as part of their routine workout program. 87% of focus group participants stated they would use a fitness or lifestyle coach in their life if the service was more affordable or easier to employ.
- Almost all focus group participants stated they would welcome an easier way to manage their health and wellness data in their life and a way to integrate all the data was needed.
- 27% of participants stated they would use information to better manage their health and wellness, but also the health and wellness of other family members.
- 100% of the focus group participants stated that they would be interested in a product that is sustainable, durable, and reliable in all aspects of operation.

TABLE 11.7

Physiological Parameters Important to Focus Group Participants

Important Fitness and Wellness Parameters

Heart rate
Skin temperature
Blood oxygen level
Blood pressure
Sleep patterns/stages
Diet and eating habits
Respiration rate
Glucose level
Electrodermal activity
Cortisol level

11.5.3 Market Review of Similar Products

Susan's team performed a market analysis to better understand the fitness and wellness customer bases and to understand readiness of company to launch the new product. Figure 11.X contains a high-level summary of the preliminary market review.

Figure 11.X   Preliminary Market Reviews for the Wearable Technology Project

Purpose of Analysis

This market analysis was performed to understand the market and its potential reediness for the Grand XI personal coach and to understand the competition base. Successful expansion of technology in this manner will significantly increase cash flow and funding returned to our investors. The result of the market analysis is documented in the report.

Industry Status—General State of Industry

The current industry is expanding product offering in the fitness health care arena. However, none of the offerings provide the level of health coaching that is expected to be achieved by the *My Saturi*.

Supply chain complications resulting from pent up demands and unavailability of critical workers to produce and transport supplies.

Potential Customer Base

The potential customer base includes the following:

- Males and females between the ages of 20–65
- Customers having an interest in improving and maintaining their health

- Potential customers are located worldwide with the exception of countries during time of certain political unrest and war, or resource (e.g., economic) limited

Customer Habits And Financial Tolerance

- Educated customers are more interested in the technology and have the potential to pay for the product.
- Fitness experts are interested in having the technology to assist their clients in achieving optimal health.
- Medical professionals have a high interest in the product and interested in potentially using in the practice.
- The customer is willing to pay between $100--$200 for the product.

Competitors

Competitors are those companies that are currently invested in and producing health-based equipment that are capable of collecting some of the data type that the Health Coach is collecting and providing to their customers. Many of these companies have a strong design capability and may be able to duplicate in term the technology being introduced by the Health Coach.

11.6   STEP 2: DEFINE—DESIGN CRITERIA AND CONCEPTUAL VISION FOR *MY SATURI* PROJECT

As depicted in Figure 11.2 the second step in applying the 4-D Algorithm for New Product Design is to *define* design factors identified in Step 1 to identify final design criteria which can be used in the design of the new application, and in the development of a product vision that is accepted and approved for prototype development and testing. Through the use of vision concepting, Art and his team incorporated their design lenses into brainstorming and the use of analytical tools, such as the political, economic, sustainable, technology, environmental, and legal (PESTEL) analysis and a strengths, weakness, opportunities, and threat (SWOT) analysis to identify change factors which could impact design criteria and product vision scenarios. It should be noted that the examples in this case study present results from both the PESTEL and SWOT analyses, but both tools do not need to be used in order to incorporate vision concepting into the design process.

11.6.1   IMPLEMENTING THE A&D VISION CONCEPTING PROCESS FOR *MY SATURI* PROJECT

By applying the A&D Vision Concepting process and design lens, as identified in Figure 11.3, Art's team identified change factors which could impact the product design criteria and product vision and scenarios. Through the use of design lens, the design factors were further evaluated by various tools, such as brainstorming,

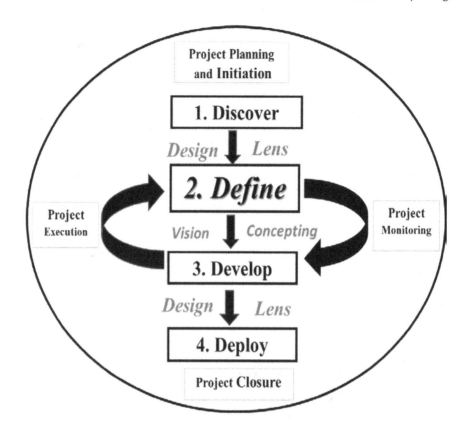

FIGURE 11.2 Step Two of the 4-D Algorithm for New Product Design.

PESTEL, or SWOT analysis. Some of the design lens guiding questions used for the *My Saturi* Project are listed below. These lenses are intended to increase the possibility of designing for inclusivity and addressing the needs of a wider customer pool. These lenses are intended for usage with Steps 2 and 3 in the 4-D Algorithm Model.

Design Lens Questions for the My Saturi Project

- What are the cultural norms for potential customers?
- What technology can be used to ensure optimal performance for all customers skin pigmentation?
- Are there medical factors or conditions (associated with a country's ethnicity) that should be considered when developing the medical panel?
- Are there any traditions that will prevent any one group from utilizing the technology from either of the three countries?
- Are there any physiological challenges for users?
- Can the product serve all race equally?
- Is the product affordable to the populace?
- Are there any language barriers?

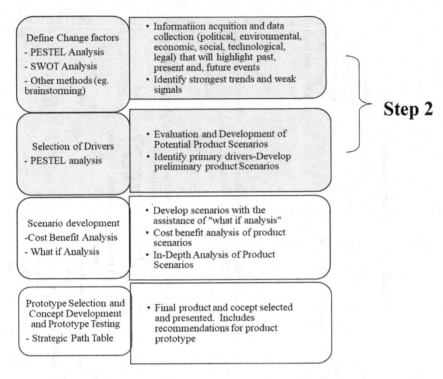

FIGURE 11.3 A&D Vision Concepting Process Components Step 2.

- Does the user require knowledge that is not easily accessible?
- Will the product in its current configuration allow usage for people with various disabilities?

11.6.2   USE OF THE PESTEL ANALYSIS FOR THE *MY SATURI* PROJECT

The design factors from Step 1 were further evaluated through use of a PESTEL analysis to identify factors which could influence or change design parameters and scenario development. The design lens identified above were considered and incorporated into the analysis by Art's team. Results of the PESTEL analysis are summarized in Table 11.8 and identified the following change factors that should be considered during the **develop** step of the 4-D Algorithm for New Product Design:

- Data integrity and trust of the data currently have legal considerations both from a political and societal perspective that will need to be incorporated into the app and marketing. This was identified as a critical design criterion.
- Cultural norms and promotion of health consciousness varies among regional locations and societies within the respective countries. This should be considered during marketing of the final product, Step 4 of the model.

TABLE 11.8

Strategic Planning—PESTEL Analysis for My Saturi Project

Design Factor	Political	Economical	Societal	Technological	Environmental	Legal
Data Integrity and Trust	All three of the countries have policies related to data management. How the data is managed may be a political issue if elevated in the future (ethical).	The cost to integrate and link the medical data to medical software will need to be explored.	There are pockets of trust issues related to electronic data and how it will be used—may be regionally defined.	All three of the countries have limited Wi-Fi in several regions. Technological expertise varies.	No identified impacts	All three countries have existing legal requirements associated with data integrity and security.
Ease of Use	No identified impacts	Cost may be elevated for customers that may need training to gain proficiency of product understanding and use	Some clients may experience difficulty because of language barriers.	Some clients may experience difficulty because of language barriers.	No identified impacts	No identified impacts
Health Consciousness	Improved management of health is politically promoted and should be considered when marketing the product(s)	No identified impacts	Level of health consciousness varies by region in each country. Cultural norms and different ethnicity may vary by region of each country and should be considered when designing the app.	All three countries allow health data to be collected, but currently not linked to medical records. Consumers have not been provided opportunity to integrate and link their physical data to their medical records. Technology must take into account different skin pigmentations and other health related factors.	Trend is towards being environmentally friendly and organic (pockets)	No identified impacts

Readability	No identified impacts	App should be programmed in multiple languages and phrases. Solar or self-generating power source must be sufficient to provide adequate lighting and visibility.	Current slang and cultural phrases need to be programmed in and easily understood. Incorporate social routines or graphics into the audio and visual displays	Most common screen projection is vertically viewed.	No identified impacts	No identified impacts
Cost	Trade agreements and tax tariffs should be evaluated for political significance on marketing.	May be driven by inflation; economic forecasts should be evaluated and factored into cost point of product.	Social media forums should be evaluated and negative feedback on cost should be considered. Cultural views on viability cost, among population classes, should be considered.	Components of the wearable technology should be cost-effective and readily available.	Components used in wearable technology should be easily disposed. Manufacturing of the wearable technology should be environmentally friendly.	No identified impacts
Sustainable and Durable	No identified impacts	Disposal of components of the wearable technology should be economically viable.	No identified impacts	The charging capability of the wearable technology should have a five-year expectancy. Components should be manufactured using latest technology.	Components of the wearable technology should be environmentally friendly and easily disposed. Components of the wearable technology should be easily disposed.	No identified impacts

(Continued)

TABLE 11.8
Continued

Design Factor	Political	Economical	Societal	Technological	Environmental	Legal
Comfort	No identified impacts	Wearable technology should be designed to minimize production costs while maximizing comfort.	The wearable technology must be designed to be operational on different body parts.	App should be readily identified at most comfortable location of the wearable technology.	No identified impacts	Legal copyrights and patents should be evaluated for aspects of the wearable technology and comfort design(s).
Data Security and Proprietary Aspects	Political views on data security and proprietary information should be investigated for potential app implications.	Cost to execute data security and proprietary issues associated with the app should be identified. Viability of the ability to secure data should be one of the highest priorities.	Cultural norms and views on medical data need to be identified and factored into the design and marketability of the app.	Ability to develop software which is technologically secure is a critical design component in order to link to medical software systems.	No identified impacts	Laws associated with data security and proprietary aspects of each country needs to be identified and impacts identified to overall project schedule.
Regulations	Political views (all prominent parties) on apps and technology should be explored to identify potential future impacts and scenarios.	Impending legislation that would impact the design or marketing of the app or wearable technology should be evaluated.	Research is needed to understand impending regulations that could be influenced by societal factors.	Research is needed to identify current and future regulations associated with technology and data management and understand impacts.	No identified impacts	No identified impacts

- Need to fully understand the technological and legal capabilities to link medical software to the new Grand XI app. This was identified as a critical design criterion for both design and execution of the project.
- App must correctly perform across all races and be easy to use. This should be promoted during design and marketing of the product and was identified as a design criterion.
- The wearable technology must be able to be worn on different locations on the body.
- App should program medical terminology but also cultural phrases and slang (modified by country). This was identified as a design criterion.
- First-of-Kind costs need to be fully understood, with adequate margin, including overall cost footprint for the manufacturing and disposal of the product. An economic analysis is needed to ensure the selected product design meets company expectations for product cost and viability. This was identified as an action for project management.
- Wearable technology must be designed to be self-charging or rechargeable on an infrequent basis, waterproof, and easily read in extreme moisture and heat. This was identified as a design criterion.
- Security of the data, especially linked to online medical software, must achieve the highest security rank. This was identified as a critical design criterion.
- Political and societal view on the legal ramifications of medical data needs to be explored and addressed. This was identified as an action for project management and the marketing team.

11.6.3 Use of the SWOT Analysis for the *My Saturi* Project

Art's design team performed a SWOT analysis to identify strengths, weaknesses, opportunities, and threats associated with the My Saturi Project. Figure 11.4 summarizes results of the analysis. The data collected was evaluated, analyzed, and catalogued for usage in decision-making with regards to the current product status and future considerations; the SWOT analysis will increase the efficiencies and the probability of evaluating the data in a manner that will provide optimum results that can be used to make a decision on product viability.

Results of the SWOT Analysis identified:

- Development of an electronic application to link personal medical data, real time, to personal medical data is novel and does not exist in the marketplace.
- It may take several years of development to achieve the change.
- Patents and taxation should be factored into development costs.

The change factors discovered through the data review were further defined in the PESTEL and SWOT analyses, represented key factors to be incorporated into the design criteria and potential product scenarios evaluated in Step 3, **develop**, of the 4-D Algorithm for New Product Design.

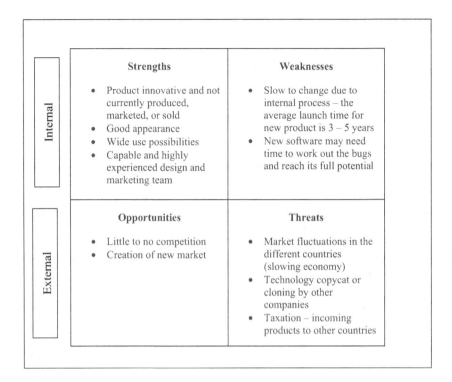

FIGURE 11.4 SWOT Analysis of My Saturi Project and Grand XI Company.

11.6.4 IDENTIFICATION OF FINAL DESIGN CRITERIA AND PRODUCT SCENARIOS FOR *MY SATURI* PROJECT

Art and his project team worked together and created not only design criteria that were associated with technological considerations, but also cultural and other factors which would have relevance to product functionality. Below are the final design criteria carried forward for *My Saturi* Project:

- Electronic application must be capable of performing artificial intelligence (AI).
- Electronic application must have the ability to integrate data from multiple sources and communicate with medical software used for data management.
- Electronic application must have the ability to link to online medical systems and download physiological data to linked medical accounts. This includes understanding the technological and legal capabilities to link medical software to the new Grand XI app.
- Electronic application must do minimum functions of the current Grand XI app (e.g., date, time, step count, etc.).
- Electronic application must make recommendations on a variety of health topics, such as food, mental health, sleeping patterns, physiological data (as identified in our survey or focus groups).

- Electronic application must have the ability to alert wearer of health trends and areas of stress and can also send the signal to linked medical accounts.
- Electronic application must consistently perform no matter race or ethnicity.
- The wearable technology must be able to be worn on different locations on the body.
- Electronic application should be programmed using medical terminology but also cultural phrases and slang (modified by country).
- The wearable technology must be designed to be self-charging or rechargeable on an infrequent basis, waterproof, and easily read in extreme moisture and heat.
- Security of the data, especially linked to online medical software, must achieve the highest cybersecurity level.
- First-of-Kind costs need to be fully understood, with adequate margin, including overall cost footprint for the manufacture and disposal of the final product selected for prototype testing.
- An economic analysis is needed to ensure the selected product design meets company expectations for over time product cost and viability.
- Political and societal views on the legal ramifications of medical data needs to be explored and addressed prior to final product selection, along with recommendations on how to minimize project risks.

The *My Saturi* project team identified three potential product placement locations to be further evaluated, in Step 3, **develop**, of the 4-D Algorithm for New Product Design for use with their new electronic application: on the wrist, around the neck, and any other place on the body (i.e., arm).

11.7 STEP 3: DEVELOP—FINAL PRODUCT SELECTION AND PROTOTYPE TESTING

By applying the A&D vision concepting process as identified in Figure 11.6, Art's team further evaluated the product placement locations from Step 2 to identify a final product for testing and corporate approval. The team used the What-If analysis to further identify strengths and weaknesses and a cost analysis of each product placement location. As depicted in Figure 11.5, the third step in applying the 4-D Algorithm for New Product Design is to define the final product and perform prototype testing.

11.7.1 WHAT-IF ANALYSIS FOR *MY SATURI* PRODUCT PLACEMENT LOCATIONS

Through the use of vision concepting and the application of design lens, Art and his design team evaluated three product placement locations for performing a What-If analysis.

- Electronic application and case would be worn on the wrist.
- Electronic application and case would be worn on the neck.
- Electronic application and case would be worn directly (attached) on the body.

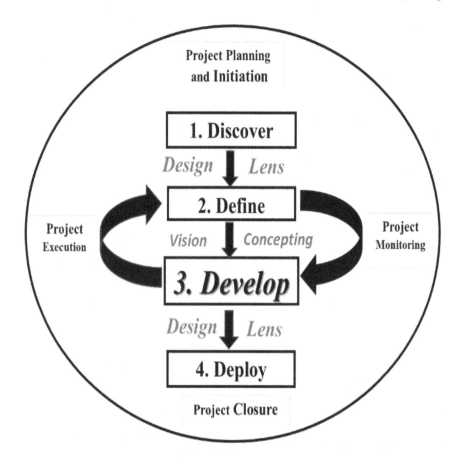

FIGURE 11.5 Step Three of the 4-D Algorithm for New Product Design.

Table 11.9 presents an excerpt of the What-If analysis performed for *My Saturi* Project and potential product placement locations. The electronic application was not analyzed in the What-If analysis because the application is universally accepted; there were no negative consequences associated with the technology, and it will not vary depending upon the population.

Results of the What-If analysis identified the following:

- Product placement location around the neck was identified as the least favorable location to wear the technology. Different types of material should be available so that it would minimize any type of allergic reaction. Any allergic reactions (e.g., latex or heat rash) could be visible to everyone and could be embarrassing.
- Wearing of the technology around the neck was least comfortable and could interfere with work activities, more so than with wearing the technology

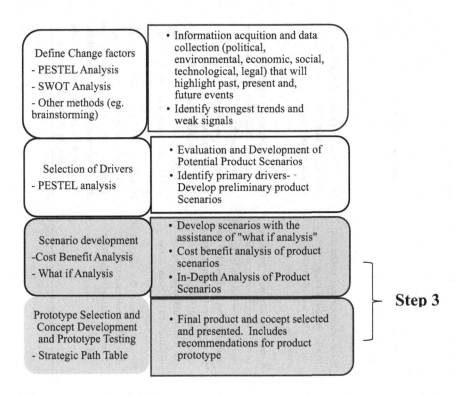

FIGURE 11.6 Vision Concepting Application for Step 3.

around the wrist or other body location. The product placement location should offer the flexibility for users to wear the technology at three different locations on the body; this enables the user to select the most comfortable location for their lifestyle. Consider offering a design of wearable technology that can be worn either on the wrist or adhering to any place on the body. Implanting of any type of monitoring device under the skin would require the use of a licensed physician in the United States and is not desirable for the average consumer.

• Physiological data can be collected from either the wrist or other body loca-tions, but it was identified that two medical data parameters could not be collected when the technology is worn on the neck. Pursue wearable tech-nology on wrist and body placement options.

Overall conclusion of the What-If analysis determined that most desirable product placement location would be technology that could be worn anywhere on the body, with the second most desirable location to wear the technology on the wrist. Art and his team then performed a cost analysis of both product placement locations to better understand if there were economic factors that would drive the decision on the final product placement location.

TABLE 11.9

What-If Analysis for *My Saturi* Project

What-If Question	Wrist	Neck	Directly On The Body	Recommendation
	Consequences	Consequences	Consequences	
What if customers are allergic to certain types of material (e.g., latex, nickel)?	Consequences would be an allergic reaction.	Consequences would be an allergic reaction.	Some customers may not be able to affix or adhere the device to the skin because of an allergic reaction.	Different types of material should be available so that it would minimize any type of allergic reaction.
What if wearing the technology is not comfortable?	The current standard for placement of wearable technology is on the wrist. There is only a small population of people (< 4%) who will not wear anything around their wrist.	A significant number of people in the population (< 35%) do not like to wear anything around their neck. In particular, around the neck could interfere with safely performing work.	A small segment of the population (<1%) currently wears medical technology on their body (i.e., arm, side), but feedback has been very positive. Implanting of any type of monitoring device under the skin would require the use of a licensed physician in the United States.	The product placement location should offer the flexibility for users to wear the technology at three different locations on the body; enables the user to select the most comfortable location for their lifestyle. Consider offering a design of wearable technology that can be worn either on the wrist or elsewhere on the body.
What if wearing the technology interferes with physiological data collection?	Current technology for wearable fitness data collection is worn on wrist. Research indicates all desired data can be collected from the wrist.	Research indicates there are two medical parameters that cannot be measured from around the neck.	Research indicates that all medical data can be collected from different locations on the body (e.g., arm).	Physiological data can be collected from either the wrist or other body location. Pursue wearable technology on wrist and body placement options. Conduct further research on the two medical parameters that cannot be measured on the neck.

11.7.2   *My Saturi* Product to Market Cost and Final Product Selection

Two potential product placement locations were analyzed for costs to build the wearable technology. Costs associated with the electronic application were not included in the cost analysis because they are fixed and the same no matter where the technology is worn. Table 11.10 contains a small excerpt of the overall cost analysis for *My Saturi* project product placement scenarios. The full cost analysis identified costs associated with each individual component needed to manufacture the product placement location. Results of the analysis identified a significant cost difference to manufacture wearable technology that can be worn anywhere on the body vs. the more traditional wearable technology which is worn on the wrist. The additional costs were identified for electronic components needed to obtain data measurements using an implant or directly adhering the device to the skin. Adhering the device to your skin resulted in lower production costs than a device that was implanted to the skin.

The cost difference for both the wrist and other body location (e.g., arm) was not significant enough to restrict development of a product that could be worn on multiple locations on the body and provides options to the consumer.

11.7.3   Prototype Development and Testing

Art and his design team analyzed all the design parameters and selected wearable technology that could be worn on the wrist, but with a feature that the electronic application can also be adhered directly to the body if desired. The final prototype selected for product testing was designed and assembled using two different types of adhesives for application to the body. Prototype testing was implemented in all three countries, the United States, Canada, and Japan, within the Gen XI customer base, but also at malls and other pop-up testing events. Overall feedback from the testing were positive with personnel asking to keep the prototypes for personal use. An excerpt of the results from the prototype testing included

- Approximately 95% of the testing population preferred to wear the device on their wrist or other part of the body.
- Wearing the device around the wrist did interfere with some work or exercise activities.
- There was no problem in reading the medical parameters because they were on the smart phone.

TABLE 11.10

Summary of Product Costs for Two Product Placement Location Scenarios

Product Placement Location	Raw Materials Costs Per Device	Production Costs Per Device	Resource Costs Per Device	Total Cost per Device
Worn on the Wrist	$5.00	$2.00	$5.00	$12.00
Worn on the Body	$9.00	$5.00	$6.00	$20.00

- Less than 1% of the test population experienced discomfort when performing work while wearing the device.
- The electronic application demonstrated the ability to link with a customer's medical records.

11.7.4 Final Product Approval for *My Saturi* Project

Art and his design team presented final results of the PESTEL, SWOT, What-If analysis, and prototype testing to the Grand XI president and board, in addition to analysis of each design parameter. Included with the briefing was a recommendation on wearing the technology on the wrist, with the ability to also adhere the electronic case and application anywhere on the body. An in-depth cost analysis was presented with certified cost estimate numbers so the Board could determine the total investment needed for production and product execution and marketing. The Board approved the manufacturing of the Gen XI wearable technology, along with a defined strategy for executing the product launch.

11.8 STEP 4: DEPLOYMENT OF *MY SATURI*

As depicted in Figure 11.7, the fourth step in applying the 4-D Algorithm for New Product Design is to deploy the final product. Art and his design team received approval from the Grand XI Board to deploy the *My Saturi* wearable technology. Fundamental to deployment of the product was development of a marketing plan to identify product line goals, objectives, and cost along with development of a communication plan which targets both potential customers, but also distributors of the Grand XI equipment.

11.8.1 *My Saturi* Marketing Plan

Insights gained throughout the design phase process were used to leverage an effective marketing strategy inclusive to a target customer base. Certain features of *My Saturi*, such as the flexibility to place the product anywhere on the body would be emphasized, along with consideration of societal factors, identified through applying VEDT, were incorporated into the market strategy. Art and several of his designers met with personnel from the Grand XI marketing division to assist in development of an effective marketing strategy and plan with the clear purpose of distribution, sales, and targeted financial projections. The marketing plan was to serve as a roadmap to a successful launch of *My Saturi*. Specifically, the marketing plan for deployment of the *My Saturi* project included the following:

- Vision statement that was consistent with the overall mission of Grand XI.
- A marketing strategy that took into account the buyer's purchasing cycle, incorporation of vision concepting and design lens, identified in Steps 2 and 3, that would be tailored by country and branding that would also identify *My Saturi* as the next generation of electronic fitness coaches. In addition, the 4Ps for marketing the product in each country was solidified: product, price, place, and promotion.

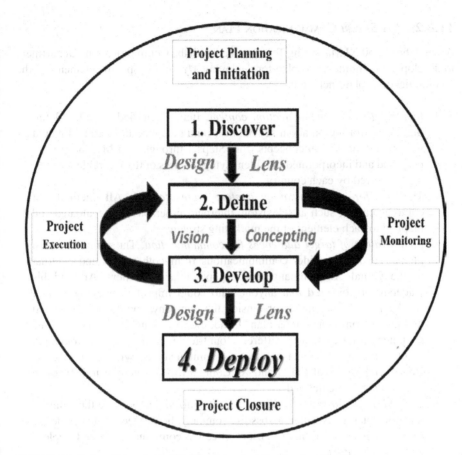

FIGURE 11.7 Step Four of the 4-D Algorithm for New Product Design.

- Manufacturing and distribution contracts. An additional marketing analysis was performed to further identify the most –cost-effective manner by which to produce *My Saturi* for both the electronic and wearable technology, along with additional distribution contracts were obtained within Canada and Japan.
- Marketing and sales goals were identified for each country. The goals included development of a strategy to minimize long-term losses and net profit for over the next 10 years taking into account identified economic weaknesses in each country. Sales targets were identified.
- Summary of financial forecasts, break-even analysis, and financial statements, inclusive of cost impacts, was developed.
- Identification of roles and responsibilities for execution of the *My Saturi* marketing strategy and plan.

Once the marketing plan had been solidified, Art and several designers assisted in development of a communication plan to ensure the success of *My Saturi*.

11.8.2 *My Saturi* Communication Plan

Art and the Grand XI Marketing Team worked with the Communications Department to develop a communications plan for deploying *My Saturi*. Specific elements of the communications plan included:

- *Identification of communication content.* Items identified included defining what information about *My Saturi* would be communicated and why it was important. Vision concepts from Steps 1 through 3 of My Saturi were reviewed and incorporated into identifying features of the wearable technology preferred by each country.
- *Methods for communicating content were defined.* All methods of communication, such as television, internet, in-person were evaluated and mapped to each element of the marketing strategy.
- *Identification of target audiences for communication.* The communication plan needed to consider communications from both within and external to the Grand XI Company, not just potential customers. Art and his leadership recognized that anyone who could impact the deployment of *My Saturi*, whether inside or outside the company, should be considered within the communications plan. In addition to communications focused on potential customers in different countries, with different customs and regulations, effective and continued communications with the Grand XI Board, employees of the Grand XI Company, and potential influences in each country were defined.
- *Defined roles and responsibilities.* Art recognized that because the communications plan is executed across the company, he needed to clearly define roles and responsibilities to ensure effective communication and deployment of *My Saturi*.

11.8.3 Implementation of Step Four, Deployment of *My Saturi*

Art and his team successfully executed production and deployment of *My Saturi* in the United States, Canada, and Japan. Highlights from Step 4 included

- Development and deployment of *My Saturi* and associated marketing plan into the United States, Canada, and Japan
- Establishment of quantifiable goals to monitor deployment of *My Saturi*
- Continuous and effective communication with customers, distributors, within the company, and stakeholders on how deployment of *My Saturi* has performed against defined monetary and production goals and objectives, as well as feedback from customers and distributors.

11.9 PROJECT CLOSEOUT OF *MY SATURI*

As part of overall project management, Art recognized the needed to develop a change management plan to close the *My Saturi* project and fully transition management to a new product line division. All roles and responsibilities of the *My Saturi* project team were transitioned to other functions, with the majority of the responsibilities assigned to a new product division created for *My Saturi*. Project personnel were reassigned to either a new project or to another assignment in the company. Prior to the team disbanding, Art scheduled a meeting and lessons learned from the *My Saturi* project were identified for consideration in future projects and by company management. All information from development and implementation of the *My Saturi* project were electronically archived for use in future projects.

Index

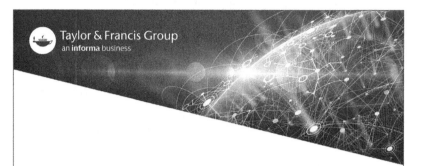

,

Printed in the United States
by Baker & Taylor Publisher Services